SEEDS FOR CHANGE

SEEDS FOR CHANGE

THE LIVES AND WORK OF
SURI AND EDDA SEHGAL

MARLY CORNELL

Des Moines, Iowa

Published by

Des Moines, Iowa
www.smsfoundation.org

ISBN 978-0-9906207-0-9 (hardcover)
ISBN 978-0-9906207-1-6 (paperback)

Library of Congress Control Number 2014951325

Cover and interior design: Mayapriya Long, Bookwrights
Maps and family trees: Elder Carson, Carson Creative

All proceeds from the sale of this book go to Sehgal Foundation.

Dedicated to future generations

SURI'S

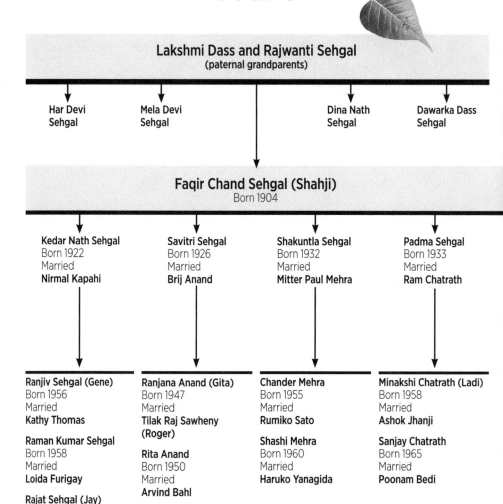

Lakshmi Dass and Rajwanti Sehgal
(paternal grandparents)

| Har Devi Sehgal | Mela Devi Sehgal | Dina Nath Sehgal | Dawarka Dass Sehgal |

Faqir Chand Sehgal (Shahji)
Born 1904

Kedar Nath Sehgal
Born 1922
Married
Nirmal Kapahi

Savitri Sehgal
Born 1926
Married
Brij Anand

Shakuntla Sehgal
Born 1932
Married
Mitter Paul Mehra

Padma Sehgal
Born 1933
Married
Ram Chatrath

Ranjiv Sehgal (Gene)
Born 1956
Married
Kathy Thomas

Raman Kumar Sehgal
Born 1958
Married
Loida Furigay

Rajat Sehgal (Jay)
Born 1960
Married
Veena Pakalapati

Sanjeev Sehgal (Raju)
Born 1966
Married
Parvinder Kaur (Meenu)

Sumeet Sehgal
Born 1971
Married
Shivani Uberoi

Ranjana Anand (Gita)
Born 1947
Married
Tilak Raj Sawheny (Roger)

Rita Anand
Born 1950
Married
Arvind Bahl

Rakesh Anand (Rick)
Born 1954
Married
Jeannette Gaylene

Chander Mehra
Born 1955
Married
Rumiko Sato

Shashi Mehra
Born 1960
Married
Haruko Yanagida

Minakshi Chatrath (Ladi)
Born 1958
Married
Ashok Jhanji

Sanjay Chatrath
Born 1965
Married
Poonam Bedi

FAMILY

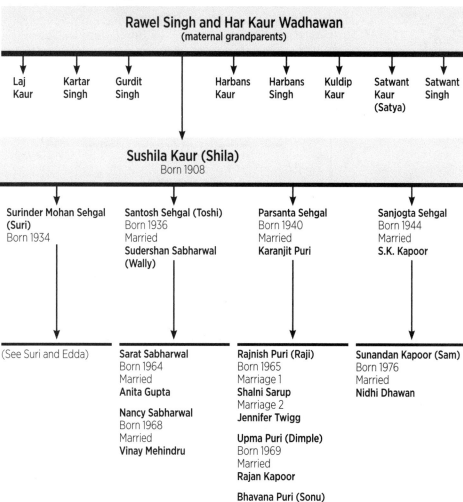

Rawel Singh and Har Kaur Wadhawan
(maternal grandparents)

| Laj Kaur | Kartar Singh | Gurdit Singh | | Harbans Kaur | Harbans Singh | Kuldip Kaur | Satwant Kaur (Satya) | Satwant Singh |

Sushila Kaur (Shila)
Born 1908

Surinder Mohan Sehgal (Suri)
Born 1934

Santosh Sehgal (Toshi)
Born 1936
Married
Sudershan Sabharwal (Wally)

Parsanta Sehgal
Born 1940
Married
Karanjit Puri

Sanjogta Sehgal
Born 1944
Married
S.K. Kapoor

(See Suri and Edda)

Sarat Sabharwal
Born 1964
Married
Anita Gupta

Nancy Sabharwal
Born 1968
Married
Vinay Mehindru

Rajnish Puri (Raji)
Born 1965
Marriage 1
Shalni Sarup
Marriage 2
Jennifer Twigg

Upma Puri (Dimple)
Born 1969
Married
Rajan Kapoor

Bhavana Puri (Sonu)
Born 1972
Married
Chad Hansen

Sunandan Kapoor (Sam)
Born 1976
Married
Nidhi Dhawan

EDDA'S FAMILY

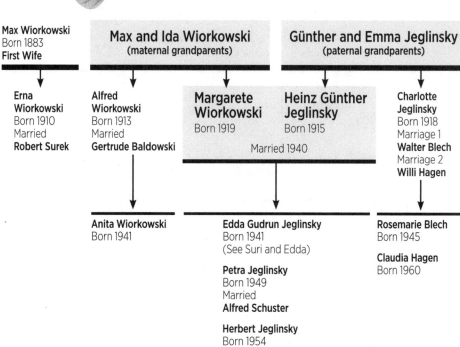

Max Wiorkowski
Born 1883
First Wife

Max and Ida Wiorkowski
(maternal grandparents)

Günther and Emma Jeglinsky
(paternal grandparents)

Erna
Wiorkowski
Born 1910
Married
Robert Surek

Alfred
Wiorkowski
Born 1913
Married
Gertrude Baldowski

Margarete
Wiorkowski
Born 1919

Heinz Günther
Jeglinsky
Born 1915

Married 1940

Charlotte
Jeglinsky
Born 1918
Marriage 1
Walter Blech
Marriage 2
Willi Hagen

Anita Wiorkowski
Born 1941

Edda Gudrun Jeglinsky
Born 1941
(See Suri and Edda)

Petra Jeglinsky
Born 1949
Married
Alfred Schuster

Herbert Jeglinsky
Born 1954
Married
Elisabeth Vels

Rosemarie Blech
Born 1945

Claudia Hagen
Born 1960

SURI AND EDDA

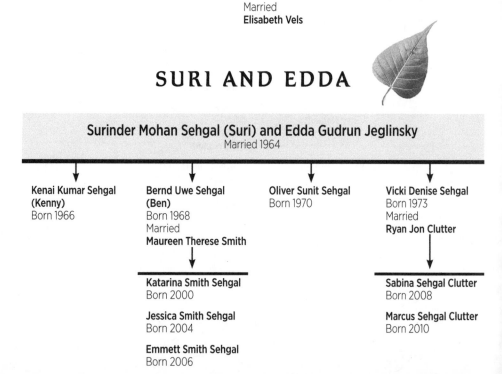

Surinder Mohan Sehgal (Suri) and Edda Gudrun Jeglinsky
Married 1964

Kenai Kumar Sehgal
(Kenny)
Born 1966

Bernd Uwe Sehgal
(Ben)
Born 1968
Married
Maureen Therese Smith

Oliver Sunit Sehgal
Born 1970

Vicki Denise Sehgal
Born 1973
Married
Ryan Jon Clutter

Katarina Smith Sehgal
Born 2000

Jessica Smith Sehgal
Born 2004

Emmett Smith Sehgal
Born 2006

Sabina Sehgal Clutter
Born 2008

Marcus Sehgal Clutter
Born 2010

Contents

Foreword

Seeds for Change: The Life Work of Suri and Edda Sehgal is an inspiring addition to an already remarkable Iowa-India agricultural and humanitarian heritage that includes Professor George Washington Carver providing nutritional advice to Mahatma Gandhi during India's quest for independence, and Nobel Peace Prize laureate Dr. Norman Borlaug bringing "miracle wheat" to India to inaugurate the Green Revolution and save millions from starvation.

Beyond its historical context, this volume is a stirring, only-in-America tale of two individuals who struggled as refugees in their home countries—one in Asia and the other in Europe—then came to the US and not only fell in love with each other but, improbably and highly appropriately, with the state of Iowa, a place with an international reputation for welcoming refugees and feeding the world.

Suri and Edda were married in Des Moines, raised their family in Urbandale, and, along the way, made contributions to a great scientific and humanitarian legacy around the globe. Inspired by Gandhi as a young boy, Suri became an admirer of legendary Iowa hybrid corn breeder and US Vice President Henry A. Wallace, who, in turn, had been mentored by Professor Carver. Wallace founded the company that became Pioneer Hi-Bred International, where Suri spent decades building its acclaimed global research and production network.

Suri had a fascinating connection to Dr. Borlaug as well, in that both of them were assisted during the early stages of their careers by a Rockefeller Foundation grant that involved Harvard Professor Paul Mangelsdorf. It was Mangelsdorf who facilitated Borlaug's initial work in Mexico, and encouraged Suri to apply to Harvard where he earned his PhD. It was in Cambridge that Suri began dating Edda, an au pair living with the family of a political science professor at Harvard named Henry Kissinger, on whose staff I later served at the White House.

While Marly Cornell's book richly details Suri and Edda's odyssey, my own introduction to the full impact of Suri and Edda's work and the Sehgal family's legacy came only in 2010 during my initial trip to

India. While there, I learned that Suri is revered as an iconic figure in Indian agriculture, both for the extremely significant contribution that he personally made to the development of the country's seed and agribusiness industry, as well as for the most impressive programs that the Sehgal Foundation is so commendably implementing in villages in the Mewat region, one of the poorest areas of India.

My wife Le Son and I were received with extraordinary courtesy and hospitality during our visit to the headquarters of the Sehgal Foundation development organization outside of Delhi, to learn firsthand about the steps Suri and Edda are taking to address this very difficult but critically important challenge to alleviate poverty. We are thrilled that, as a result of that visit, each summer an Iowa high school student spends eight weeks working at the Sehgal Foundation as a World Food Prize Borlaug-Ruan intern, further enhancing the Iowa-India connection.

We were also greatly impressed by the magnificently energy-efficient building that the Sehgal Foundation constructed as the headquarters for its operations in India, which had attained LEED Platinum certification. Suri and Edda's commitment to sustainability impelled me to redouble my own efforts to reach that same highest possible level of energy efficiency for our World Food Prize Hall of Laureates building in Des Moines, Iowa. With my desire to emulate their example and the Sehgal Family Foundation's generous support, we too succeeded in reaching this goal.

In looking back at the end of this volume, an aspect of Suri's and Edda's lives that has particular resonance for me is how, not unlike my own family, they always seem to find themselves caught up in some of the worst natural and man-made disasters: the Nazi depredations during World War II, the Partition of India, political revolution in the Dominican Republic, and an earthquake in Japan. At home on the Iowa prairie or in the Punjab, the Sehgals have always persevered. Whatever the obstacle, through dedication, diligence, and sheer determination, they have brought great success to everything they undertook and, as it is clear, shared their blessings widely with others.

Ambassador Kenneth M. Quinn, president
The World Food Prize Foundation
Des Moines, Iowa

Introduction

International entrepreneur, humanitarian, and philanthropist Dr. Surinder (Suri) Sehgal is a principal elder among a distinguished lineage of experts in world agriculture,[1] who became over the course of his career one of the primary players in the development and worldwide dissemination of high-quality hybrid seed—earning a rightful place alongside eminent crop scientists, seedsmen, and agricultural visionaries Henry A. Wallace, Paul Mangelsdorf, and William Lacey Brown. How Suri became a respected and pivotal figure in the development of the global hybrid-seed industry and, in partnership with his wife Edda, how they used their combined skills and good fortune to make a positive difference in the lives of many of the poorest people in the world can best be understood by taking a brief glimpse back in time.

Suri was born in 1934 in the old Punjab, which was then northwest India under the British Raj and is now Pakistan. He was the second son and fifth of eight children of a Sikh mother and a Hindu father in a land-rich but, by then, cash-poor family. The Partition of India in 1947, dividing British India into Pakistan and India, brought sudden outbreaks of violence in Punjab, Bengal, and elsewhere in the country, shattering Suri's idyllic childhood.

When Muslims began attacking and killing Hindus and Sikhs, Suri's father's effort to send three of his daughters to the relative safety of India resulted in a desperate measure—he pushed thirteen-year-old Suri onto a refugee train to escort and protect a younger sister who had become separated from the others in the commotion. Wearing only a shirt and shorts, and sandals that he would lose along the way, Suri spent the next two months helping his sisters reach a family friend, then going in search of his uncle, armed only with the knowledge that his uncle worked "somewhere near Delhi."

---

1. *Maize Genetics and Breeding in the 20th Century,* ed. Peter Peterson and Angelo Bianchi, World Scientific Publishing Company, 1999.

Homeless and penniless in Delhi, Suri ate whatever scraps he could find in refugee camps and slept in empty train cars. He witnessed the same horrific violence and brutality he saw before leaving home, but now it was Sikhs and Hindus carrying out revenge killings against Muslims. He sought comfort at the public prayer grounds, listening to Mahatma Gandhi speak of peace and address the hatred that was ripping apart India. Suri found inspiration in a stirring speech by Prime Minister Jawaharlal Nehru about the need for harmony among India's different peoples. The roots of Suri's lifelong commitment to make a difference in the world were deeply implanted during this period.

Once his family reunited in Amritsar, India, in severely reduced circumstances, Suri completed high school and college (studying biology with a major in cytogenetics) and formulated a plan to attend graduate school in the US. Suri wrote to Harvard professor and renowned evolutionary biologist Paul C. Mangelsdorf, a giant in the field of genetics and an expert in the origin and evolution of maize. Mangelsdorf answered with an invitation for Suri to apply for graduate work at Harvard University.

While at Harvard, a series of serendipitous encounters and collaborations brought Suri together with other great thinkers, scientists, and visionaries of the day and gave him the opportunity to take the first step in forging a career that would eventually span the globe. Suri thrived in the collegial atmosphere of intelligence and high spirits where everyone studied and worked hard.

Suri's life changed forever when friends introduced him to a recent arrival to the US, a beautiful young German woman who was learning English and living as an au pair in the home of Harvard professor Henry Kissinger and his wife Ann. Edda Jeglinsky came from a background as turbulent as Suri's. As a small child she had been evacuated with her family from German Silesia in January 1945, just ahead of the onslaught of the Soviet army. Her family's escape from danger eerily paralleled Suri's. They were both children of displaced cultures, whose elders had lost their deep roots in ancient ways of life. Together in the US, Suri and Edda not only found each other but also an altogether new and different world where their destiny would result in the fulfillment of their version of the American Dream and helping others fulfill their dreams.

Suri and Edda raised four children and made it possible for other relatives to immigrate to and/or study in the US and abroad. They continue to make profound differences in many lives, not only within their families and the rings upon rings of others radiating out from them, but also with many in the international community who have been helped through their philanthropic organization in India and in the US, the Sehgal Foundation.

CHAPTER 1

Five Waters

Suri's earliest childhood memories are blissful ones. He was one of six or seven little schoolboys sitting cross-legged in the shade of a sprawling peepal tree in summer just before monsoon season. As his teacher reviewed arithmetic numerals and the Urdu alphabet, Suri gazed at the sunlight through the heart-shaped leaves of the sacred tree, and his eyes followed the seemingly endless roots that wound down from the branches and up from the earth, wrapping around the wide trunk. Everything around Suri looked beautiful to him.[2]

School was always let out by noon in such hot weather. In the afternoon, Suri ran to the courtyard of his family's *kothi* and climbed the jamun fruit trees that grew all along the boundary walls. The tall trees were loaded with sweet black plums that turned Suri's tongue purple. At night the boy and his cousins kept cool sleeping in the courtyard under the starry sky.

Our story begins here in the old British Punjab Province of colonial India that is now part of Pakistan. Centuries ago, the region was named after the five rivers that run through the fertile territory of the exotic Indian subcontinent. In Persian, *panj* means "five," and *ab* means "waters." The Punjab was the land, Punjabi the spoken language. This

---

2. The peepal or pipal, *ficus religiosa,* also known as Bo-tree or Bodhi tree and considered sacred to Hindus and Buddhists, was the tree the Buddha was meditating under when he achieved enlightenment. Mahatma Gandhi deemed it a sacred symbol of the unity of all religions. The tree produces oxygen day and night, and its leaves, bark, roots, and juice are known for their many medicinal purposes.

agriculturally rich terrain was about to be severed at the cusp of two nations in a tragic partition brokered by the departing British Raj.

In the early nineteenth century, Suri Sehgal's paternal great grandparents sold their properties in Bhera, a small predominantly Muslim town where they lived on the banks of the Jhelum River. As Hindus, the Sehgals were in the minority in Bhera and were aware of tensions between the different religious communities. So the family moved east-northeast about 127 kilometers and put down roots in Guliana, about 15 kilometers north of Lalamusa.

The town of Lalamusa was the main railway junction in the Gujrat district, on the principal north-south railroad line connecting Peshawar in the north with the old capital city of Lahore, the great heart of the Punjab. The small village of Guliana was a business hub for the surrounding area and home to a diverse population of Hindus and Muslims who lived in relative harmony. Trade in the region was dependent upon agriculture, and the two communities celebrated the changes in the agricultural year as one people, even participating in each other's religious festivals. In general, the land in this region was owned by Hindus who lived near the center of town. The fields were more often cultivated by Muslims on a sharecropping basis. They lived near their fields.

The Sehgals purchased land, set up shops, and engaged in a flourishing business, trading agricultural commodities. They financed farmers in return for their crops in a traditional landlord-tenant relationship. The winter staple crops included wheat, barley, black grams similar to small garbanzo beans, mustard, and oilseed rape. The staple crops during monsoon season (July to September) were pearl millet, also called *bajra* and used primarily for fodder, and sorghum called *jowar*. At the end of each growing season, the spring (*rabi*) harvest and the autumn (*kharif*) harvest, the family bought back the grain crops, stored them, then sent them on to Lalamusa, where they were bagged, bundled, and shipped to other parts of India.

Guliana had a main street, not paved in those days, with about a dozen shops. By the end of the nineteenth century, all of them were owned by the Sehgal family. The post office was housed in a tiny shop rented from the Sehgals. They owned several farms in the surrounding villages, as well as ten thousand acres of forestland on the outskirts of town, which they leased to the Forest Department. Separated from the village by a seasonal creek, the forests flooded heavily during monsoon season, recharging the aquifers. The water table was high, making water easily reachable for drawing with a hand pump for most of the year.

The family built a Hindu temple in Guliana. Suri's grandfather, Lakshmi Dass, named after the popular Hindu goddess of wealth and prosperity, wore only white clothes as a sign of his spiritual devotion. Every six months, he made a pilgrimage to the Vaishno Devi temple in Kashmir, a shrine dedicated to a manifestation of the mother goddess.

Lakshmi Dass and his wife Rajwanti shared a commitment to helping the poor. Whenever the monsoon failed, they gave money to less-fortunate families to dig wells. If they had financed farmers for their crops and the crops failed, the unpaid loans were deferred without penalty or even written off entirely. Promissory notes written in Urdu and dating from the 1890s, which were never repaid, are evidence of the family's benevolence.

The Sehgal family belonged to the Kshatriya caste, called Khatris. In the ancient traditional social hierarchy of the Hindu caste system, Khatris were rulers and property owners who served as warriors during conflict and were responsible for governing during peacetime. They were bound by scripture to rule with justice. The Sehgals felt a sincere responsibility for protecting and helping those less fortunate and assisted actively in managing the affairs of their village. However, whether these traits were connected to their Kshatriya heritage was not a topic of conversation within the family. The expectation of service toward others was assumed as the way to be.

Though the rigid caste system had mellowed somewhat over many centuries, it still had great influence on occupations and social customs, particularly marriage. Marrying outside of one's caste was not acceptable, and all marriages were arranged by the elders in the family. Sehgal family marriages were no exception.

As the family expanded, they lived as one large extended Hindu household, with thirty-five to forty people occupying three homes next to each other on a side street. None of the houses had indoor plumbing or running water. A hand-operated water pump was in an outside courtyard of each house. Parents, grandparents, uncles, aunts, and cousins lived, ate, and worked together. Several cooks and servants were hired by the family. The diet was strictly vegetarian. No alcohol or smoking was permitted.

Less than a kilometer from the village was a cattle barn, where cows and buffalo were kept that supplied the family's dairy needs. All food for the family came from their own farms—cereals from the fields, and fruits and vegetables from around the family's kothi on the outskirts of the village near the seasonal creek.

The kothi was a palatial structure, a spacious residential building in the center of a sizeable compound. This large bungalow served also as another living space for any overflow of houseguests and as a cool and pleasant refuge during the hot summer months. A large-capacity water pump and rooms for bathing were on one side.

In one corner, several small huts were reserved for *sadhus*, holy men who had renounced all material possessions and desires of the world to spend their lives in pure devotion to God. These yogis, monks, and ascetics often wandered throughout the country and survived by accepting offerings of food. The Sehgal family shared their bounty readily with sadhus, welcoming them as honored guests. The holy men came often to the kothi and stayed for weeks or months at a time.

Considerable space in the vast yard was devoted to vegetable gardens and flowers—marigolds, roses, and jasmine. Along the boundary walls of the kothi were several dozen jamun fruit trees that served as a wind barrier. The nutrient-rich fruit and its seeds had multiple medicinal properties that were used by Ayurvedic and Unani physicians in treating various ailments, wounds, and diseases from diabetes to migraine headaches. A storage structure near the kothi, called a *godown*, held the wheat and other grains after the April/May harvest.

The kothi featured a huge finished basement with twelve prayer rooms. The prayer rooms were small, private, dark rooms with no adornments. In the morning, male members of the family went there to exercise, bathe, say morning prayers, and meditate. Men sat cross-legged and closed their eyes. Sometimes they could be heard chanting a simple mantra for an hour or so at a time.

Those coming by train from other cities to the Sehgals' home would get off in Lalamusa and from there travel the fifteen kilometers between Lalamusa and Guliana on horseback or bicycle along a single dirt road. Visitors and villagers could rent horses from men who then walked alongside them to and from the railroad stations. In some instances, a two-wheeled horse-drawn carriage called a *tonga* was used. However, this mode of transportation was less frequently possible because the dirt roads were not maintained in those days. The Sehgals owned three horses and a tonga. Their horses were kept busy each day fetching relatives and friends to and from the train station in Lalamusa.

Because the temperatures from May to June were often over 40°C (104°F), water was needed during any travel. To provide drinking water for thirsty travelers, the family had wells dug along the dirt trail between the villages of Guliana and Arah and also along the dirt road between Guliana and Lalamusa.

Lakshmi Dass and Rajwanti had five surviving children: two daughters and three sons. Suri's father was the first male child to survive. Prior to his birth, his distraught mother consulted with a holy man for help and advice. She had two daughters already, but her male babies had each died at birth. The holy man advised Rajwanti to pledge to give away her next son to a *faqir*, an ascetic holy man, to assure that the child would live. When Suri's father was born in 1904,[3] he was as pledged given to a faqir, just as his mother had been directed to do. Because Rajwanti kept her agreement, the parents were then permitted to adopt back their newborn son. For this reason, the baby was given the name Faqir Chand, his first name recalling his adoption from the holy/poor man. Everyone called him Shahji as he came of age: *shah*, because he was, in fact, from a wealthy family and not poor like a faqir, with the honorific "*ji*" added. This name was how he was known throughout his life.

Shahji was nine years old when his father fell quite ill. Lakshmi Dass was a diabetic in a time when there was no known allopathic treatment for this disease, and there were no physicians in their town. The

---

3. The birthdate is noted within the family as May 22, 1904, but there is no official record of the birth.

nearest doctor was summoned. Sardar Rawel Singh,[4] who lived about five kilometers away in the small village of Arah, was a medical officer in the British army and home on leave when he received word that he was needed in Guliana. The physician rode on horseback over the dirt trail between the villages to care for Lakshmi Dass.

The doctor was a distinguished Sikh gentleman, more than six feet tall. He had a beard and long hair that he wore under a turban. His full name was Rawel Singh Wadhawan. The surname Wadhawan was rarely used; most people in the Punjab routinely used first and middle names only. The middle name, Singh (meaning "lion"), was given to all male members of Sikh families.

A common bond united the two patriarchs at their very first meeting. Lakshmi Dass and Rawel Singh were both devoted philanthropists, serving their communities as they could with their special skills and resources. Though somewhat different in religious faith and practices—one family Hindu, the other Sikh—each man recognized in the other a deeply shared ethic and philosophy of service to others.

So close in values were the men that a decision was made the same day to establish formal family relations. When Lakshmi Dass offered to pay for the medical visit, the doctor refused. Instead, the men agreed to the engagement of a son and a daughter when the future couple came of age. Marriage between Hindus and Sikhs was not at all uncommon. The Sikh doctor's daughter Sushila, called Shila, was five years old when she was promised that day to the Hindu businessman's nine-year-old son, Shahji. The two men shook hands on it.

Nine years later, at ages fourteen and eighteen, Shila and Shahji were married, and Shila moved to the Sehgal family compound. Honoring both religions, she went on a regular basis to the Hindu temple the Sehgal family built and to her family's *gurdwara* (Sikh place of worship) when in Arah.

Before he died, relatively young in his early fifties, Lakshmi Dass called together his three sons to tell them that he was leaving each one an equal portion of his estate. His daughters were both married by then, and they had already each received a generous dowry (daughters in those days had no expectation of further inheritance from their

---

4. Sardar is the honorific preceding the name of male Sikhs, similar to "Mister."

parents). Lakshmi Dass was a successful entrepreneur who had started many businesses. He told his sons that if they managed their portions carefully, their wealth could last for generations. But he advised them to stick to their roots and do only what they knew best how to do. He specifically cautioned them not to seek business outside the local region.

Neither Shahji nor his youngest brother, Dawarka Dass, heeded their father's advice. Soon after Lakshmi Dass died, the brothers used the family fortune to engage in business ventures and investments outside Guliana and Lalamusa. The middle brother, Dina Nath, who suffered from diabetes like his father, stayed behind to tend to the family business.

Optimistic and ambitious, Shahji and Dawarka Dass went into the risky commodity futures market and an insurance business in Lyallpur (now Faisalabad) and Gujrat. Shahji took carefully considered ventures into the stock market. But sadly, this was absolutely the wrong time for business of this kind. The Great Depression that began in America had also reached British India, and the Punjab as well. The Sehgal brothers suffered great losses. The family's cash flow dried up. The brothers moved back to the family homes in Guliana with little money and pressing liabilities. They still owned many fixed assets in the village—houses and lands—but there was no income and no prospect of renewal any time soon.

Shahji and his brothers decided to split up the remaining assets they held. The land and homes were divided among the families, and the brothers went their separate ways in business.

Shahji Sehgal was a man of great principles in business and in his personal life. Meticulous in appearance, he was always clean-shaven and wore freshly ironed clothes. Known and respected as scrupulously honest, he lived by the same high ethical standards and commitment to justice and service that his parents emulated.

While raising his family in Guliana throughout the 1920s, Shahji traveled frequently on business to Lyallpur, and to the capital city Lahore, which was a straight shot 172 kilometers south by train. A multicultural center, Lahore was the site of an ancient university as well as beautiful mosques, temples, and gardens. The cosmopolitan metropolis was the seat of the Punjab government and a center of higher learning, with affiliated institutes all over the Punjab. Though Shahji had not had a formal education (his family had engaged a teacher to tutor him at home), he understood the value of education. He became an active member of the Sanatan Dharm Sabha, a committee of Hindus that reformed educational institutions. He was involved in building schools and providing aid to various institutes and orphanages. His strong interest in justice eventually led him into politics.

Shahji joined the Indian National Congress party where he came into regular contact with its principal leaders, including Mahatma Gandhi (Mohandas Gandhi was most often called *Gandhiji* with affection and respect); his trusted lieutenant, Jawaharlal Nehru; and other important educators and activists of the period, such as Pandit Madan Mohan Malaviya who made popular the slogan, "*Satyameva Jayathe*" or "Truth alone will win."[5] Shahji was strongly moved by speeches made by these men in favor of political reform and freedom from British rule in India.

Shila was traveling with Shahji when they participated in a political gathering in the industrial city of Kanpur on the banks of the Ganges River, where Gandhiji urged his countrymen to boycott cloth imported from England and cease buying or wearing any mill-made cloth. The British were buying cotton from India at low cost, which was used to manufacture high-priced clothing to sell back to Indians. When Gandhiji suggested that everyone, instead, wear handspun clothes, Shahji complied immediately. He burned all of his Western clothes. From that day on, he wore only clothing made in India, called *khadhi*.[6] Most Congress party members did the same, including Nehru, of course, who was previously known for his elegantly tailored Savile Row suits. Handspun clothing and a Gandhi cap became the uniform of the independence movement. And Khadhi, now with a capital "K," became associated with the ideology of the freedom movement.

As a result of Shahji's attendance at these political meetings and his increased involvement and interactions with the Congress, he became the general secretary of the Congress at Lyallpur and helped to organize the communities of Guliana and Lalamusa. He was often accompanied at political meetings by his younger brothers and his best friend, Amar

5. http://www.kamat.com/kalranga/freedom/malaviya.htm. Pandit means learned scholar.
6. Khadhi, or khaddar, is usually cotton/hemp, sometimes silk or wool.

Nath Sharma, from Palampur. Shahji's uncle, Gokal Chand (Lakshmi Dass's younger brother), was involved in the independence movement as well. They aligned their activities and efforts with Gandhiji's goal of his country's freedom in keeping with a steadfast commitment to peace and harmony among the diverse religions of the region.

At the Indian National Congress in January 1930 in the capital city of Lahore, the *Purna Swaraj* declaration was made in favor of complete Indian self-rule apart from the British Empire. Gandhiji encouraged the use of nonviolent civil disobedience campaigns that often ended in his arrest and the incarceration of other Congress leaders as well. When a twenty-three-year-old revolutionary freedom fighter name Bhagat Singh was hanged in 1931, many of the Congress leaders, including Shahji and his uncle, Gokal Chand, were put in jail for a short time.[7]

By this time, Shahji and Shila were well on their way to a large family. Their first child, a boy named Kedar Nath, born in 1922, was followed by three daughters: Savitri in 1926, Shakuntla in 1932, and Padma in 1933.

The economic effects of the Depression had affected India profoundly. In May of 1934, with India still in a deep depression, a plague was spreading across the country and the weather was uncomfortably warm. April through June were the hot months that came before monsoon season. The family, including a very pregnant Shila, had moved from the village to the kothi for some relief from the oppressive heat. Surinder (Suri) Mohan Sehgal was born on May 16 in that spacious country house.

After Suri, the couple had three more daughters: Santosh in 1936, Parsanta in 1940, and Sanjogta in 1944.

Suri was contented as a lone boy with a trio of sisters on either side. He enjoyed a fairly carefree early childhood in their company. His brother Kedar, twelve years older, was away at school in Lalamusa from fifth grade on, living as the youngest resident in a dormitory. Shahji was a trustee of the school, so his son was well cared for by the warden and

---

7. At Shahji's funeral in 1978, his best friend and fellow Gandhian, Pandit Amar Nath Sharma, who after Partition became a minister of Education in the Panjab government, revealed that Shahji had been helping to support Bhagat Singh's family with Rs. 500 every month since the freedom fighter's death in 1931.

his wife. The brothers saw each other only on infrequent visits during holidays when Suri was young.

Suri seldom saw his father, either. Shahji was away sometimes for weeks or even months at a time in Lyallpur or Gujrat. When he was home in Guliana, he seemed to always be busy building and tending the family business or involved in his community work. However, Shila was a strong-willed and loving woman who held the family together when Shahji was away. The warmth of Suri's large multigenerational family, with cousins to play with, constant visitors, and multiple celebrations, made for a happy upbringing full of activity.

Holidays were important in Suri's youth. Most involved fire and sweets. *Dussehra* was always a big festival held each October after the fall harvest, celebrating the victory of good over evil as depicted in the great Hindu literary epic *Ramayana*, where the Hindu god Rama is victorious over the demon king Ravana who had kidnapped Rama's wife Sita. Everyone in town dressed in colorful clothes and gathered together around huge bonfires to watch Ravana burn in effigy. Children watched excitedly as the figure burst into flames.

The festival of lights, *Diwali*, followed quickly, continuing the victory celebration of good over evil. Suri helped to put oil and cotton wicks in earthen lamps (*diyas*) that were lighted in every home. This holiday meant firecrackers, new clothes, gifts for children, visits with grandparents, and eating lots of sweets.

Bonfires at midnight celebrated the winter solstice, *Lohri*, usually on one of the coldest nights of the year. Children dressed in their woolens and were allowed to stay up late to eat sesame candy and sing songs around the bonfire.

Bhasakhi (or *Vaisakhi*) was celebrated on April 13 when the wheat crop was ready to be harvested. Feast or famine depended on the wheat harvest in the Punjab. Everyone expressed thanks for a bountiful yield with dancing. Men performed the *bhangra*, a Punjab folk dance. Women danced as well, but not with the men. This festival was purposefully celebrated around water, often near the river, in order to acknowledge the vital importance of water, the central element of this great agricultural region.

Suri was enormously fascinated by water. He enjoyed building small channels and watching the direction and speed of the water flow. On vacations he gravitated to mountain streams, canals, and waterfalls. He appreciated all of water's complexity. Observing his early interest in such things caused his family to suspect that he would someday become a civil engineer.

The Muslim observance of Ramadan, held during the ninth month of the Islamic calendar, meant that Muslims age twelve and older did not eat or drink during daylight hours for a full month of prayer, fasting, and charity. At the end of Ramadan, a huge feast, called *Eid al-fitr*, celebrated the breaking of their fast. Muslim friends dressed in colorful outfits and brought Suri's family lovely large plates of food to share. Though Muslims were not vegetarians, they brought the Hindu Sehgals rice pilaf and other tasty vegetarian food on this important holiday.

Guliana had one primary school, a *madrassa*, for boys only. Primary school had four grades, with children starting first grade at age five. School was held Monday through Friday with a half day on Saturday. Suri's brother Kedar had attended this school until he finished fourth grade.

There was no school for girls in the village, so Suri's older sisters were sent to other towns to attend primary school. Savitri was sent to Jhelum to live with Shahji's sister, Har Devi. Shakuntla was sent to Lahore to stay with Shila's sister, Harbans Kaur. Padma was sent to Arah to live with Shila's parents where there was a very modest primary school for girls. This left Suri as the oldest child at home for a while.

Suri attended the madrassa through second grade. Teaching was done in Urdu, the state language of the Punjab. The teachers were mostly Muslims (*moulvis* or *imams*), but there was no reading from the Koran or religious teaching. Though his teachers were competent, Suri was not particularly inspired by the schoolwork. The core subjects, starting the first year, were writing, which was right to left, and counting. There were no sports, games, art, or music.

Inside the classroom, each student sat on the floor atop a spread-out piece of fabric called a *tat*. Like the homes, the schools were not heated. On dry sunny days in winter when the temperature dropped, holding classes outside in the school courtyard in the sunshine was more comfortable. The same was true for the hottest days in summer when learning took place while students sat under a shade tree.

Even as a young child, Suri was known within the family and among his friends for being active, alert, disciplined, and resourceful. If he set his mind to some activity, he never gave up until he'd done what he set out to do. His activities were varied but he especially enjoyed helping with the harvesting of vegetables and fruits each year.

Because of the different family cultures represented by his Hindu father and Sikh mother, Suri grew up in both faiths with no exclusive adherence to one or the other. Although different in some aspects, the two religions were harmonious in practice in the Sehgal home, just as they were in the Punjab. Both faiths practiced daily prayer and meditation, and both of Suri's parents embraced the concept of a single supreme God. Hindus also enjoyed statues and images of various lesser deities—gods and goddesses who each had their own super powers, but Sikhism did not recognize such deities. Though Shahji himself was a strict Hindu, assisting the less fortunate was the only true strictness he imposed upon his children, indirectly, by example. Suri was keenly aware of his family's compassion for the poor and their support for schools, orphanages, and holy men. He knew his father meditated in the kothi and that both of his parents honored the sacrifices and prayers of the holy men who came to stay there, making sure that wholesome hot meals were delivered to them during their visits.

Suri and his younger sisters never knew Shahji's parents. Lakshmi Dass had died before Suri was born. Rajwanti died while he was still a small child. Paternal grandparents in India were typically called *Dada* and *Dadi*, with the affectionate "ji" added: *Dadaji* and *Dadiji*.

Maternal grandparents in India were typically called *Nana* and *Nani*, also with the affectionate "ji" added. These were the names the children used when addressing Shila's parents, Rawel Singh and Har Kaur Wadhawan. The middle name Kaur, meaning "princess," was given to all female members of Sikh families. The Sehgal children visited their Sikh grandparents several times throughout the year and for a few weeks each summer. Shila usually accompanied them, but Shahji did not due to his work obligations.

The family trip to Arah from Guliana was always on horseback. The children were aware that their grandparents were well known in the village due to Doctor Rawel Singh's unique and generous medical practice. Soon after the end of the Great War, the doctor had taken early retirement from the military due to severe arthritis in his knees. He had to use a cane to walk. Living on a pension and settled on the family farm, he served the community by providing free medical service to the people of Arah and those in neighboring villages. For two hours every morning and two hours later each afternoon, anyone could come to his clinic, located in the family's *haveli*, similar to a kothi, for free treatment.

The Sikh doctor was highly regarded in Arah, well educated and well respected. He and Har Kaur had nine children, including Suri's mother.

Their large home was on the edge of the village, with a ground floor, a first floor, and a second floor. All the rooms for main activities were on that first floor: the doctor's study, his room for receiving visitors, the living room, dining area, kitchen, and one bedroom. Other family members had rooms on the ground floor, which was where the grandchildren and any other visitors stayed. In summer months, beds were set up in the courtyard for sleeping in the open. The top floor was an open space with a storage room on one end. Guests often slept up there in summer months if they could not be accommodated comfortably elsewhere.

All the children, including cousins from Lahore, enjoyed going to Arah to visit *Nanaji* and *Naniji*. The cousins had fun together there. Their favorite activities were climbing the trees around the haveli, hunting for mushrooms, playing cards or carom board and, in the evenings in winter season, making a bonfire in the courtyard to roast potato mini tubers or black grams (chickpeas in small pods) and feasting on the nutritious, tasty treats. Suri once fell out of a large neem tree[8] he was climbing near the haveli. Though no bones were broken, he was in bed with fever and pain for several days and cared for by his physician grandfather.

---

8. The neem (Indian lilac) was a popular tree in Ayurvedic medicine known to have marvelous healing properties for treating problems related to blood sugar and many other ailments. http://kumareshgupta.hubpages.com/hub/Neem-Tree-Indian-LilacA-miracle-herb.

In the living room of the house were several framed photographs on the wall, depicting groups of thirty or forty men in uniform. Photos like that were taken whenever Rawel Singh received some recognition or award, and the doctor was a highly decorated soldier.

Nanaji never told the grandchildren any stories about his time in the war, but *Naniji* did—even long after the Great War ended. Har Kaur had been worried sick about her husband throughout the war years 1914–1919. She thought it was a miracle that he did not die in a combat zone.

Rawel Singh had tried to assure his wife that the medical units were not on the frontlines where the casualties were, but further behind where wounded soldiers were brought for medical care. All those years later, Har Kaur still cried each time she narrated the horror stories of the war.

Their oldest son, Kartar Singh, had followed in his father's footsteps, both in becoming a physician and in joining the army. He was based in the Nowshera Cantonment in the north near Peshawar for the duration of World War II, which started when Suri was five years old.

Suri and the other children were spectators to the effects of the war close up when visiting their grandparents' village. The names of soldiers who died in combat were published in the newspaper. In addition, whenever there was a death of a soldier, the soldier's family was sent a telegram in English. Everyone in the village knew when any family received one of the dreaded telegrams. Because most of the villagers in Arah were illiterate, families with relatives in the army assembled every day at Suri's grandparents' house with the telegrams to learn the fate of their loved ones.

Suri watched his grandfather recite the names in the newspaper and translate the telegrams from English to Punjabi or Urdu for the families who wanted to know details about a family member's death. The young boy was present with the families who were devastated by their losses, and he witnessed the temporary relief on the faces of those who did not hear their loved ones' names recited.

There were sometimes ten to twelve people in the room, mostly women, since most of the men were away in the war. Suri stood nearby when his grandfather gave the bad news to those who lost loved ones. The minute the women's worst fears were confirmed, they began wailing and crying loudly. Suri often cried along with the sad mothers, wives, and sisters of men killed in the war, while the kind doctor tried to offer words of consolation to the distraught women.

Suri and his sisters adapted easily to the differences between his relatives in Arah and those in Guliana. The children observed their Sikh relatives eating meat and the adults enjoying alcoholic beverages that were strictly prohibited in their own home. The kids knew to be particularly well behaved in the company of their physician grandfather—to sit up straight and be polite. Nanaji still led a regimented life long after his retirement from the army. He was a kind man, but a strict disciplinarian and far more formal in demeanor than their father. The children showed their respect for him, as was the custom when first greeting parents or grandparents, by touching his feet, after which Nanaji would pat the child on the back and give a blessing. The grandchildren received no hugs or any other form of physical affection from Nanaji, but Naniji was a very affectionate grandmother who hugged them often.

The grandchildren kept out of the way when Nanaji first appeared early in the morning. The good-hearted doctor rose early, prepared for work, and walked with his cane to his clinic to see patients. His clinic was always full of ill or injured people waiting to see him. When Nanaji returned, by 10:00 a.m., for his breakfast, the children took off running for the haveli to play there until their grandfather left again for his late afternoon clinic hours.

The haveli was a lovely place with boundary walls on all sides and fruit trees and plenty of spaces for children to play. Family members went there in the morning to bathe and prepare for the day. Hot water was available in winter months, boiled on a *chulha* stove outside the bathing room. Inside rooms included the clinic and bedrooms for overflow guests. Two adjoining godowns (warehouses) stored enough wheat grain for the household needs for a year and *bhusa* (wheat straw) and fodder for the animals as well. Adjoining the courtyard was enough land to grow vegetables throughout the year and keep four or five milk animals (cows and buffalo) and some chickens.

The Persian well in the corner of the compound was essentially an open hole about a dozen feet wide that was dug down as deeply as needed to reach the water table. In summer the well was filled to the top; in winter the level went down. A large wooden water wheel with a chain of buckets brought up the water with the help of a team of bullocks (oxen) hitched to the central drive shaft. When the bullocks walked around, occasionally prodded by a *mali* (gardener), the wheel turned, lifting up buckets full of water. Suri was fascinated by this process and seeing how the water was then used to irrigate vegetable plots and fruit trees in the haveli. He sometimes climbed on a seat mounted behind the yoked bullocks and enjoyed a ride.

Similar to the Sehgals, the Wadhawan family was self-sufficient. Fruits, vegetables, milk, whatever they needed, came from their haveli. When guests other than strict vegetarians visited the Sikh household, which happened often, the meal served was usually goat or chicken curry. Even though they had a flock of chickens at the haveli, every now and then they ran out, and the family servants were sent to the neighboring villages to buy more.

On one occasion, when Suri was about seven, he accompanied his grandparents' servants on an expedition to buy some live chickens. They walked together to a nearby farm. In the courtyard of the farm, while three of the servants were busy taking care of the business purchase, a growling dog lunged at Suri. The large dog grabbed Suri around his waist and left a deep gash in his abdomen. The servants rushed to help Suri and bandage the wound, then they took him home. Suri was bleeding quite a bit and in terrible pain that lasted for several days. His grandfather gently treated his injury, and no infection developed. However, as a result of that frightening experience, Suri never wanted a dog as a pet.

Despite his lack of formal education, Shahji was highly competent in public affairs and within his leadership roles in the Congress Party. He was talented at mathematics and could do complicated computations in his head, and he was well versed in Urdu, the language in which business correspondence in the Punjab was done at the time. However, English was the *lingua franca* among people throughout India. Not knowing how to speak fluent English, the elder Sehgal was unable to participate in such affairs at the highest levels, which grieved him deeply. For that reason, he became increasingly adamant about seeking a good education for his children. He had tried to make his youngest brother, Dawarka Dass, go to England for higher studies. Gandhiji and Nehru had done that, and it was common practice for Indian students who could afford to do so. Shahji was disappointed when Dawarka Dass was unwilling to pursue higher studies even in Lahore.

Far more than their Hindu and Sikh religions, disparities in language and education were the main differences in the upbringing of Suri's parents. Shila knew how to read and write, but only Punjabi; whereas Shahji, proficient in Urdu, didn't know how to write in Punjabi. They could communicate verbally, but not in writing. When Shahji was away from the family, they could not maintain meaningful contact. The only telephones were in government buildings; there were none in private residences in the villages. For them to be out of communication for a few months at a time was common due to Shahji's business travel and political organizing.

Shila's brothers were well educated. They had all been sent to schools in Lahore, unlike the girls in her family who were tutored at home, mostly in the local Punjabi language. Her oldest brother, Kartar Singh, had gone to medical school. The second oldest, Gurdit Singh, was now a police inspector in Lahore. The third, Harbans Singh, also in Lahore, had a position in the supply division with the military. The fourth brother, Satwant Singh, was still finishing college, but later worked for the government as a labor inspector. At some point, they all served in respected roles within the British administration as a result of their good educations.

Shahji's focus on education became more of an obsession as his children were growing up. Like Gandhiji, Shahji was a firm believer in gender equality. In 1941 Suri's younger sister, Santosh, was ready to start primary school, but Shahji didn't want to send away any more of his daughters to live with relatives in order to be educated. As soon as Suri finished second grade that year, the family moved south to Lalamusa where there were several schools, some of which Shahji and his close friend, Amar Nath, had been instrumental in establishing.

CHAPTER 2

Lalamusa Junction

The Sehgals' spacious new two-story home was located on the road leading to the Lalamusa Railway Station. Only a few meters from the house was the Grand Trunk Road,[9] which connected Lalamusa to Kharian to the north and Gujrat to the south by almost the same distance. A police station sat at the corner of the crossroads. Military convoys were seen frequently on the Grand Trunk Road, and trains full of soldiers and guns rumbled past as World War II raged on. Whenever a train came through the Lalamusa station, Suri and his friends rushed to watch it go by. The soldiers sometimes threw packets of biscuits (cookies) to the children gathered on the platform.

Suri's uncle, Doctor Kartar Singh, traveled back and forth by train between Nowshera in the north and his reporting unit in the south. He often traveled by Frontier Mail, a fast train that connected Peshawar with Bombay (now Mumbai). When he could, he wrote letters ahead of time to Shahji, saying when his train would come through Lalamusa. Shahji and Shila went to the railway station to watch and exchange waves whenever a train he was traveling in was passing through.

---

9. The Grand Trunk (GT) Road dates back to ancient times and criss-crosses the Indian continent.

The move to Lalamusa meant that Kedar was able to move home for the first time since before Suri was born and attend the Sanatan Dharm High School for boys. Suri's school-age sisters were enrolled in either a middle school or a primary school for girls. The primary school for girls was next to the Lalamusa railway station, and the primary school for boys, where Suri enrolled in third grade, was just on the other side of the Grand Trunk Road. For the first time in many years, the entire family was living together.

Suri's school had four grades and included Hindu, Muslim, and Sikh children. Instruction was in Urdu as it had been in Guliana. Suri would retain bitter memories of his new school, because the teachers were neither skilled nor compassionate. Caning was a common practice. The teachers enforced the prevailing wisdom that this type of punishment was necessary for discipline. The only good thing about the school was that Suri could easily walk there from home.

The Sehgal family occupied the first floor of their new home. On the ground floor were four warehouses that had previously been used for storing commodities. One of the four warehouses was rented to a carpenter who specialized in making trolleys for vendors. Once in a while, he fashioned a toy out of wood for Suri. The boy was thrilled to have a miniature version of the carts with little wheels that turned. Two warehouses on the ground floor were empty. The fourth was full of cigarettes.

In the depths of the worldwide Depression, Shahji's friend Amar Nath, called Uncle by the Sehgal children, had recommended that Shahji go into the cigarette business, that it was supposed to be lucrative. Despite the family's religious prohibition of smoking, in his desperation during the Depression, Shahji had acquired a distributorship for a new brand of cigarettes for the northern zone, primarily to service the Northwest Frontier Province and Kashmir. But the business did not do well, and Shahji had to bear the loss.

Adjoining their house was a Sikh gurdwara. Next to it was an Arya-Samaj temple. A little further toward the railway station was a blacksmith (*Lohar*) shop where the blacksmith's son, Muhammad Alam Lohar, could often be heard singing beautiful Punjabi folk songs. Suri and other kids used to gather around to listen to him.[10]

Opposite the family home were several hundred acres of land that belonged to the government, arid land where nothing grew. A family-owned farm a few kilometers in the direction of Guliana was cultivated by sharecroppers. That farm supplied most of the family's food, primarily wheat and seasonal fresh vegetables.

Despite the family's previous reverses in business, the Sehgal home was still a vibrant hub for relatives coming from near and far, to and from Arah or Guliana. A day didn't seem to pass without guests showing up for one meal or another. Almost every meal turned into a celebration of some kind. Such easy association was a daily ritual within the warm and close-knit family.

Suri and his family continued to travel regularly, either to Arah, Lahore, up to a hill station (resort town) in the mountains, or a day trip to the banks of the Jhelum River for the Bhasakhi holiday. A few times they traveled back to the family homestead in Guliana where some of their cousins still lived. They stayed in the main house or in the kothi, which was still shared by everyone in the Sehgal family. The kothi was a wonderful vacation escape, especially in the summer months.

The family sometimes traveled by tonga; however, that took much longer, so they usually preferred to rent horses for any travel that was not by train. Horses and a few tongas were always available for rent at the train stations. The frequent trips to Arah to visit Suri's grandparents were by train to Kharian, the closest train station. Kharian, on the Grand Trunk Road, had become a principle military base for recruiting and training troops from the surrounding villages and towns into the army to fight in Europe, North Africa, and Southeast Asia.

For Suri and his family, the continuous backdrop of World War II remained an ever-present reality in Lalamusa as well. Soldiers, guns, and army recruiters were a regular presence. At one point, the authorities had constructed a type of defense structure between the Sehgal home and the railroad station in case of air attacks, which, thankfully, never came to Lalamusa. The structure was more like a big empty swimming pool, about an eighth of an acre in size, where Suri and the other neighborhood kids liked to play. There were no manufactured toys in those

10. Muhammad Alam Lohar recorded his first album at age thirteen and became a popular singer in Pakistan.

days per se, so kids improvised for entertainment and made up games. Suri and his friends played *gilli-danda*, a simplified version of cricket in which a stick called a danda is used to hit an oval-shaped piece of wood called a gilli, or *kabaddi*, a contact sport played in teams that was a combination of tag and wrestling.

After a circus came through Lalamusa with performing animals, acrobats on tightropes, and stunts on swings, Suri and his friends organized a circus show in a courtyard near the railway station. Unfortunately, one of the kids fell from the tightrope during the first performance. He wasn't seriously hurt, but that was the end of their circus.

Movies were sometimes shown in the open air at the railway station in Lalamusa. Suri saw *Ali Baba and the 40 Thieves* there. As part of the aggressive recruitment campaign by military authorities, short clips were shown after the movie that depicted the heroism of Indian soldiers in the war. A group of musicians and singers traveled from one city to the other by train and performed after the movie shorts were shown. One musician played the *dolkhi* (a roped two-sided drum) and another played the harmonium (a keyboard and bellows instrument). They sang the praises of life in the army. One beautifully performed Punjabi song, "*Burti Ho Jao*" became a theme song in the Punjab, which basically conveyed the message: join the army and have a good life.

Several of Suri's cousins were attracted to the idea of military service; every generation in Shila's family was represented in the military. But Shahji continued to make it clear to Suri that he would go to college one day rather than into the army.

Suri always looked forward to the half-day train ride to Lahore to attend family weddings or spend an extended vacation with cousins in the big city of Lahore. Wedding trips usually meant a one- to three-week celebration staying with family. Two of Shila's married sisters lived in Lahore. Her oldest sister, Laj Kaur, lived on Ram Gali (lane of God), where her husband, Jagat Singh Thapar, was a successful publisher of high-quality educational magazines in Urdu that were distributed to Indian schools. The other sister, Harbans Kaur, lived on the outskirts, and her husband, also Sikh, had a prominent position in the government.

Suri and two of his cousins of a similar age used to pal around together in Lahore, sometimes getting into mischief. During one visit for a family wedding when Suri was eight, he went out with his cousins late one afternoon to play around a nearby canal. The water was flowing rapidly and the boys hung far out over the rushing stream. Suri leaned too far toward the water while holding onto an embankment and lost his balance. He was about to be swept away in the fast flowing water and easily could have drowned, but his cousins grabbed him and pulled him to safety.

That was not the first time Suri almost drowned while enjoying the magic of water. Only a few years before, when visiting another aunt on the outskirts of Lahore, he had gone with a different cousin to a reservoir and had to be pulled out of the water there as well, when he'd leaned out the same way.

Suri and his cousins went to movie theaters and sat in the wooden chairs on the sloped floor similar to a school auditorium. The balconies were fancier with padded seats. A huge attraction in Lahore was the Anarkali Bazaar.[11] The renowned boutique marketplace had been attracting people from all over the Punjab and neighboring states since the 1600s. Pungent aromas and endless sounds and sights could be found in and around the rows of small shops. Tailors were ready on the spot to measure and sew colorful embroidered garments. Every type of confection was offered by sweet-makers (*halvai*) who sat by open fires, stirring milk and sugar to make candies with flour and nuts, spices, carrots, and syrups. Local sweets in winter were also made with pumpkins, squashes, and sesame seeds. The smell of spicy kabobs and samosas filled the air. In those days, each vendor was only a few feet from the next, sitting on the floors of their shops side by side along narrow walking paths, selling their snacks, crafts, shoes, clothes with decorative stitching, and the all-important jewelry—a must for dowries. Silver, gold, and glass bangle bracelets came in every color and texture.

Suri and his cousins visited the snack shops there in the afternoon for *pakoras* (veggies deep-fried in batter) and local sweets. Two of Suri's favorites were *burfee*: a sweet confection made from condensed milk, sugar, and nuts that was served cold; and *jalebi*: a crisp treat for special occasions such as birthdays, festivals, and weddings. The boys inhaled

11. Anarkali means pomegranate blossom in Persian.

the scent and watched as the halvai deep-fried a batter made with flour, yogurt, sugar, and spices, and then soaked it in sugar syrup right before serving the jalebis—hot and delicious.

A visit with Suri's cousins in Lahore during kite festival season resulted in an unpleasant altercation. Vasant (or Basant) Panchami, a minor holiday held on the fifth day of February, marked the beginning of spring. Kite flying was a huge sport at that time. Suri and his cousin with almost the same-sounding name, Sohinder (so the family referred to him as Lahoria Sohinder to differentiate the two), were flying their kites when a neighborhood kid intentionally cut the strings of their kites while they were floating high in the air. The kites blew away and were never found.

Suri and his cousin grabbed the boy and punched and slapped him until he started crying and ran home. The boys were now scared that they were in trouble. They ran home and stayed out of sight, fully expecting a visit from the boy's parents. But no one came that night.

As a proactive step, Suri and his cousin were up and out early the next morning. As they walked along the street, they gave a formal greeting to each adult they saw, folding their hands, bowing slightly, and saying respectfully, "*Namaste.*"

People in the street smiled approvingly. A couple of folks even complimented the boys on their good behavior and manners. Later that day, a neighbor showed up at the house and told the parents how well behaved their boys were. Suri and his cousin were delighted, mostly because no one ever showed up to complain about their having punched the boy who cut their kite strings.

Throughout World War II, the British had continued to recruit heavily from India. Before the war ended, more than two-and-a-half-million Indians, the largest volunteer army in history, had served, and tens of thousands had died. The majority of soldiers from the Punjab were Muslims, but all groups were represented in large numbers. However, service to the British cause in World War II was not necessarily an endorsement of the British Raj in India. The worldwide Depression had left so many families in financial distress that jobs with the military were often the only realistic options for many in the Punjab and throughout India. The independence movement was building in strength, and to have so many Indians fighting for "the sake of freedom," when they themselves were not free while living under British rule, was an inherent contradiction noted by many, including Jawaharlal Nehru.

Since everyone on the Sikh side of Suri's family was aligned in some way with the military or the government, the Wadhawans were essentially on the opposite end of the political spectrum from the Sehgals, who were solidly nationalistic and fully participating in the campaign against the British occupation. However, despite this difference, the families were warmly connected and respectful of each others' views.

Though the Wadhawans were aware that the Sehgal home was a meeting place for those participating in Congress party activities, politics was not a topic of conversation when the two families were together. To have such strongly activist relatives might have reflected negatively upon them as government employees, but the most Shila's brothers would do to try to discourage the Sehgals' political organizing was to sometimes suggest that such activities were probably a waste of time.

Shahji remained deeply involved in his political work and now served as district president of the Congress party. The Sehgal home was a bustling center for community organizing.

In August 1942 Mahatma Gandhi planned to launch a large civil disobedience campaign called the Quit India Movement that he knew would result in many arrests of Congress Party leaders. Gandhiji wrote a personal letter to Shahji beforehand, asking him not to participate in the campaign. Gandhiji knew that Shahji had been suffering for a couple of years from poor health, including a brief paralysis of his legs when on a trip to Lahore in 1938. He had recovered with the help of Ayurvedic and Unani medical practitioners, but his physical movements were still sometimes slow and painful. Gandhiji did not want Shahji to risk his health any further by being arrested or spending any time in jail as a result of civil disobedience campaigns. Shahji, of course, complied with Gandhiji's request. As predicted, most of the Congress leaders were arrested immediately after Gandhiji's speech on August 8, and the beloved leader remained in prison, along with thousands of his followers, for the duration of the war.

Kedar became the first Sehgal to complete a formal high school education. Shahji did not push his son to go abroad to study further due to his fragile health. Kedar suffered from severe asthma throughout his childhood, and there was no effective treatment for the condition in those days. He was bright, but his final grade in high school was not outstanding; he completed matriculation in the second division earlier that year at the age of twenty. Shahji wanted Kedar to pursue college in Lahore, but Kedar had no real interest in higher studies. He was anxious to get out on his own and took a job in Palampur right after high school.

Unfortunately, there was no high school for girls in Lalamusa by the time Savitri finished eighth grade. Because a closer eye was kept on adolescent girls, they were not sent away to school, so Savitri instead helped out more at home.

By the time Suri finished primary school in 1943 new kinds of enmity and religious tensions were already seeping into communities within the Punjab as a result of an increased emphasis on having different schools for children of different religions. Government schools were secular, but they were located primarily in big cities like Lahore. There were none in small towns or villages. All private education in the Punjab beyond primary school was now typically divided on the basis of religion. For those next years, grades five to ten, Hindu boys attended the Sanatan Dharm High School, Sikh boys attended the Khalsa High School, and Muslim boys attended the Islamiya High School. Schools for girls were similarly divided, although many of the teachers in all the schools were Muslim.

These divisions had the devastating effect of making religion, caste, and class serve as counters to a long tradition of communal peace in the Punjab. When taking the train to visit relatives in Lahore, Suri found it strange that hawkers sold Hindu tea and Muslim tea. The two teas were identical. Corresponding distinctions were practiced with food. In all social areas, separation was slowly becoming the norm.

As in many of the smaller villages, the majority of the people in urban areas of the Punjab were Hindu and Sikh, while those living on the outskirts of town were predominantly Muslim. There were far more Muslims in general in the Punjab, and some villages were all Muslim. Hindus and Sikhs were often better off financially, and they owned more of the land. With increased attention to these disparities, and so much emphasis on separation, jealousy and animosity amongst Muslims against Hindus and Sikhs began to escalate, and vice versa.

When nine-year-old Suri started fifth grade at Sanatan Dharm High School on the outskirts of town, he had to walk through empty fields and along the road up to the main street more than three kilometers each way. Temperatures ranged between 38° and 45°C (100°–113°F) in the summer months, making Suri miserable on his walk home each afternoon in the hot sun.

Classes in Suri's school were still taught in Urdu. English was a new subject for him. The school had only one playground for field hockey and soccer. Suri loved playing both; he loved sports. However, with the school so far from home, there was no way to play there in the evenings. Instead, in the huge empty lot in front of his house, Suri played games with the children of neighboring railway employees. He sometimes made friends by giving out free cigarettes from Shahji's large stockpile in the family warehouse.

A couple of times, out of curiosity, Suri tried to smoke, but he never did develop a taste for tobacco. He knew he would get into big trouble if he ever came home with his clothes smelling of cigarette smoke.

The summer after Suri turned ten, in 1944, the family went on a holiday trip to Palampur, where Suri's brother Kedar was now working in a civilian role with military supplies near the hill station there. Palampur was an important military station. Some British officers and business people had permanent residences there. The resorts in the mountains were often crowded during the summer vacation periods.

Monsoon rains continued incessantly for days and days that year. The roads were heavily damaged and the bridges washed away. As a result, the Sehgal family was happily stranded in Palampur for a longer period than expected.

Suri loved all the water. He took full advantage of the rain. He collected rainwater, built aqueducts, and watched to see how the water flowed from one place to another.

The family was away from home so long that the school year began, so Suri was admitted to the local Roman Catholic mission school in Palampur, St. Paul's Senior Secondary School, located in a beautiful pine forest. This unexpected change in plans turned out to be fortunate, because the mission school was so much better than Suri's school at home. He felt truly challenged in school for the first time. He became interested in his studies and enjoyed learning. He loved the teachers, and they recognized him as a promising student.

Suri was befriended by a small monkey in Palampur. She had shown up one day, climbing down from a tree, and Suri fed her and played with her. He named her Projna. The two bonded and shared a sweet affection. Projna sat on Suri's shoulder and accompanied him as he hiked around the hills.

A big attraction during this respite was a military parade of the Gurkhas, the most decorated British regiment. They were mountain people from Nepal, short in stature, but tough fighters known for their loyalty and bravery. Suri and his family joined the large crowd that could always be counted on at military cantonments whenever there was a significant victory for the allied forces during the war. As hundreds of Gurkhas marched in the parade, led by a band, Suri thought they looked very impressive in their crisp uniforms, with shiny boots, and feathers in their hats.

Shahji was able to do some traveling back and forth on his own while the family remained in Palampur. But finally, after several months, the roads were open and regular travel was possible, so the family prepared to go home. Suri was deeply disappointed to leave such a stimulating school situation, and to have to say good-bye to his friend Projna.

Though sad to return to the Sanatan Dharm School in Lalamusa, Suri was resigned to the fact that there was nothing he could do about it. He had no respect for the cruel teachers and no interest in the studies. But at least he now knew what a truly good school could be like. At the end of each academic year, he did well in his exam and was promoted to the next grade with little effort.

Suri was eleven when World War II finally ended. The news was everywhere. Over the next weeks, soldiers were returning to their villages. The family received a letter saying that Suri's physician uncle,

Kartar Singh, was headed home by way of a train passing through Lalamusa. Shila and the children went to the train station to wave and cheer her brother's train.

After the devastating loss of tens of thousands of Indian soldiers during the long years of World War II, the time had come for celebration as many families throughout the Punjab were reunited. Suri and the other children gathered excitedly for the military victory parades.

The Indian Congress Party leaders who had been incarcerated were released, including Mahatma Gandhi and Jawaharlal Nehru who had by then spent three years in prison due to their civil disobedience campaigns against the British. This event was received with great jubilation all over the country, in Lalamusa, and within the Sehgal home.

By the time the war ended, the sentiment in Britain had finally shifted in the direction of liberating India from colonial control. Though they were on the winning side of the war, Britain was economically compromised. The Labor Party was in power. Jawaharlal Nehru was now the principal negotiator for India's freedom, along with Mahatma Gandhi and the All India Congress Committee, which included Sikh, Muslim, and Hindu leaders.

In the midst of the postwar excitement and the coming independence, the Sehgal family celebrated the marriage of Suri's oldest sister in 1946. Savitri had barely turned twenty when she married Brij Anand on May 16, Suri's twelfth birthday. The groom had been a classmate of Kedar's in high school, and Savitri knew of him as well before the marriage was arranged.

The eating, singing, and dancing festivities surrounding Savitri's wedding lasted for a week. Friends and relatives came from all over the Punjab, at least one or two members of each extended family. A friend had a kothi nearby with a beautiful yard and garden full of fragrant flowers, three floors, and plenty of extra rooms. Between the Sehgals' main house and their friend's kothi, there was enough room to accommodate all the out-of-town relatives and guests.

Preparations began several days before the wedding—lots of activity with one celebration after another. Relatives were everywhere, and the atmosphere was full of cheerful interactions, music, and all manner of foods and sweets. Female relatives helped to prepare the bride. On the morning of the actual wedding, they applied lovely mehndi (henna) designs on Savitri's hands and forearms. Suri thought his sister looked quite beautiful when he saw her in her wedding clothes and jewelry.

His sister's arranged wedding was the first time Suri gave much thought to the place of women in Indian society. He began to notice the differences in the choices and confining roles in life that his sisters lived within. From that point on, he had a growing awareness of gender inequities. His father's and Gandhiji's lifelong commitment to social justice, decency, and fairness was not lost on Suri.

CHAPTER 3

The Midnight Hour

Freedom was in the air. There was plenty to feel good about. India's independence was within sight. However, not all leaders in the All India Congress Committee had the same goal. In August 1946 the head of the Muslim League, Mohammad Ali Jinnah, with the support of a majority of Indian Muslims, publicly demanded the partition of India and the formation of Muslim majority areas into a homeland to be called Pakistan.

Just before starting eighth grade that fall, Suri organized a soccer team of his friends, including Hindus, Muslims, and Sikhs. The players got along well despite the growing friction elsewhere in the Punjab, which they were hearing about with greater regularity. Though most people who lived in the villages near Lalamusa during this period remembered that time as free from interpersonal or religious conflict, violence was gathering in cities far away.

By January 1947 occasional outbreaks of hostility occurred in some parts of the city of Lahore. An intermittent sense of urgency was looming. Suri's maternal grandparents decided to leave the Punjab for an extended visit with their son, Gurdit Singh, whose job had just been transferred from Lahore to a town near Delhi. Shila's parents planned only to take a break from the tension, which they fully expected would be temporary.

The month of February brought the exciting birth of Shahji and Shila's first grandchild. Suri was now an uncle. Savitri and Brij Anand had a baby girl they named Ranjana. Their little family lived 2.4 kilo-meters away in downtown Lalamusa.

While the Sehgal family was enjoying the newest member of the family, the movement toward independence suddenly accelerated. The British Labor government announced the decision to officially end British rule in India by 1948. British Lord Louis Mountbatten was appointed viceroy in March 1947 to broker the independence process. At the same time, riots had slowly spread to other cities in the Punjab, including the Rawalpindi district in the far north. By the time Suri fin-ished eighth grade, the Partition saga had intensified dramatically, and the (now) thirteen-year-old was immersed in a dangerous and brutal chapter of history that was about to play out, literally before his eyes.

In June 1947 Lord Mountbatten announced the Indian Independence Act. The British viceroy's plan included moving up the date of secession of the Indian Union from the British Commonwealth to August, only two months away, and designating the creation of two new independent dominions, Pakistan and India, and defining the partition of the provinces of Bengal and Punjab, with exact boundaries "to be determined."

The Punjab would be partitioned between the two free nations, one Muslim and one Hindu and Sikh. The dividing line would separate the West Punjab in Pakistan from East Punjab in India. Nehru and Gandhiji had worked tirelessly to keep India united; but in the end, Nehru agreed to the partition in spite of Gandhiji's protests.

The difficult agreement would leave large pockets of Sikhs and Hindus in the predominantly Muslim Pakistan. The same was true in reverse for parts of India that were predominantly Hindu with large pockets of Muslims. However, despite the disparities, there was no apparent expectation of trouble with the transfer of power from Britain to the new governing bodies of each of the newly liberated countries. Pakistan was to come into being on August 14, and the "new" India would be free at the stroke of midnight on August 15.

Many of the Hindus and Sikhs, including the Sehgals, in the western area of the Punjab that would be in Pakistan accepted the new reality without too much concern. Though outbursts of Hindu-Muslim violence were increasingly common throughout the rest of the country, citizens in the Punjab had long prided themselves on living in peace despite religious differences. Even the sporadic attacks on Hindus and Sikhs were mostly looked upon as merely one more isolated riot.

In celebration and preparation for the coming independence, the freedom movement, which still included Hindus, Sikhs, and Muslims working together, began having public meetings in the Lalamusa town square. But by July, violence had spread closer to home, in Lahore and Amritsar.

People in Lalamusa were aware of disturbances elsewhere, but their town had remained relatively calm. Few people were concerned to the point that they considered relocation. The Sehgals had many close Muslim friends and neighbors. Shahji felt confident that his family was not in any real danger.

The monsoon was late that year, no rains yet, and the temperatures were soaring to 43°C (110°F) during the day. A few days before August 14, "freedom day," as the end of Ramadan was nearing, some Hindus and Sikhs were wounded in a neighboring village and brought to the medical dispensary near the Lalamusa police station across the street from the Sehgals' house.

Suri and Shahji were two of only six or seven people there to help the injured, doing whatever they could with medical supplies and bandages. Young Suri was shocked to see so much bleeding and so much suffering in this one small place. He looked to his father and followed his example, doing quickly whatever anyone asked him to do amid the chaos and cries of pain—one thing after the other—lifting or holding people still. The Muslim doctor was trying to treat each of the wounded people, some of whom had missing limbs. Blood was everywhere. A few of the injured did not survive. This was Suri's first glimpse of dead people.

Feeling scared and wondering what was going on as he watched the doctor work, the boy smelled danger, not knowing what else might happen. But once Suri and his father had done all they could do to help that day, Shahji dismissed the dreadful incident as an unfortunate isolated event. The elder Sehgal could not imagine anything getting in the way of the exciting changes to come with their country's freedom, only days away.

Early on the morning of August 14, 1947, the local freedom fighters held a flag hoisting at the railway station in Lalamusa. The atmosphere was festive. The British Union Jack was brought down, and the new flag of Pakistan was raised. Suri and his family and their friends celebrated enthusiastically. Shahji had been invited to speak with other dignitaries at the ceremony. Hindu and Muslim leaders preached the message of communal peace and unity. Idealism was sincerely reflected in the happy crowd as Shahji spoke that morning.

The family returned home from the flag-raising celebration in a cheerful mood. One of Suri's uncles, Surinder Singh Lamba (married to Shila's younger sister, Kuldip Kaur), was staying nearby. He joined the family for a hearty lunch followed by a card game, a family tradition on holidays. Card games were a regular pastime in a room near the top of the stairs.

Suri's father and uncle, along with eighteen-year-old Gurbaksh Singh (pronounced Ger-bucksh Sing), the son of a Sikh priest from the gurdwara next door, recruited Suri to be the fourth in a card game of sweep. Suri, even at thirteen, was already recognized as a competent sweep player. Gurbaksh Singh was a promising young man the Sehgals had helped by financing his high school education in Lalamusa. His father, as a priest, had no resources to send Gurbaksh to school, and the priest's family was very grateful.

The four enjoyed one another's company and engaged in a spirited card game that afternoon. But late in the day, someone outside yelled that smoke was rising near the edge of downtown. Everyone dropped his cards, and the family went up to the roof to see what they could. The gurdwara near Suri's high school had been torched.

The family's first concern was for the safety of a Sikh priest and his family who lived in the gurdwara compound. They worried as well for Suri's sister Savitri, her husband Brij Anand, and their six-month-old daughter, who lived only a short distance from the gurdwara in the center of downtown. The Sehgals and their visitors kept a tense vigil on the rooftop until nightfall. They could hear the commotion as people shouted and shops were looted. There was no way of knowing if their friends and family were safe. And no helpful news was to come that evening.

Everyone was scared. The Hindu police sub-inspector, a trusted acquaintance, came to the house. He said he expected trouble and feared that he could not defend his community against an attack. He had only a dozen policemen. Eleven were Muslim; one was Sikh. The sub-inspector invited the Sehgal family to spend the night inside the police station grounds.

On this same evening, about 600 kilometers away in New Delhi, at one hour before midnight, Jawaharlal Nehru gave his first speech in the Constituent Assembly in his new role as the prime minister of the new India, in honor of the attainment of India's independence. His historic speech began, "At the stroke of the midnight hour, when the world sleeps, India will awake to life and freedom. A moment comes, which comes but rarely in history, when we step out from the old to the new, when an age ends, and when the soul of a nation, long suppressed, finds utterance. It is fitting that at this solemn moment, we take the pledge of dedication to the service of India and her people and to the still larger cause of humanity."[12]

His moving speech went on, promoting peace and celebrating freedom. His words were met with continuing cheers from the surrounding crowd. Mahatma Gandhi was not present at the huge celebration. Strongly opposed from the beginning to any partition of India, he had begun a fast in an effort to curb the escalating violence and promote religious harmony.

Back in Lalamusa, the sub-inspector, his son, and the Sikh police officer stayed on guard throughout the night to defend the police station. The turmoil in the city went on the entire night. The sounds of continuing upheaval could be heard in the distance. The Sehgals stayed inside the police station, but no one slept that night.

Early the next day, Shahji left to see what he could find out about what had happened. From someone who had run away from the violence, he learned that the downtown Sikh priest and his family had

12. "A Tryst With Destiny" speech by Prime Minister Jawaharlal Nehru was delivered to the Constituent Assembly of India in New Delhi on August 14, 1947. *The Guardian: Great Speeches of the 20th Century.* http://www.theguardian.com/theguardian/2007/may/01/greatspeeches.

been killed by "criminal Muslim elements" who initiated the upheaval. Shock had reverberated through the center of town as Hindus and Sikhs sought refuge. Townspeople fled their homes, taking only cash or gold, trying to find safety in the larger houses or in the Hindu temple.

The Sehgals returned to their rooftop perch and continued to watch what was going on in the distance along with several friends and neighbors. Fear mounted as they saw a huge crowd assembling at the far southwest end of the empty field in front of their house. Suddenly, hundreds of Muslims carrying spears, swords, and daggers began screaming and running toward the Sehgals' house.

The Muslim policemen fired only into the air, unwilling to shoot at other Muslims. The Sikh policeman, the sub-inspector, and his son were the only ones prepared to defend the police station. Their fear rising, the Sehgal family members raced through their home, gathering kerosene, oil, acid, and whatever defensive weapons they could find. They didn't know what else to do. They had no guns, nothing to defend themselves from an angry mob.

If the violent mob had attacked, there would likely have been a massive slaughter. But the horde rushed past, in the direction of the city. From their rooftop perch, the Sehgal family's terror shifted again to fear for the safety of Savitri, Brij Anand, and the baby.

The mob began running faster toward the city, shouting *"Allahu Akbar!"* (God is great!) At that moment, a military convoy arrived, apparently responding to a call from the Lalamusa police sub-inspector to the military station in Kharian, thirteen kilometers away. The police station was the only place in town with a telephone, other than the post office. As the convoy arrived, they saw the screaming throng, quickly stopped in front of the police station, and took defensive action.

The soldiers were mostly Hindu, but the officer in charge was British. He spoke only in English, shouting orders to his men. Soldiers entered the Sehgal home, raced up the stairs to the flat roof, and set up their machine guns. Some found other high points in town. Immediately positioned and clearly intent on defending the town, the soldiers began firing on the mob.

Suri stood to the left of the British commander on the Sehgal rooftop. The officer allowed the boy to stay close to him, watching what was happening as he fired down on the throng. People were hit by the bullets, and Suri saw them fall, some injured, some probably dead. The mob had not expected guns. Right away, the crowd started to retreat, dragging away their fallen comrades. The city of Lalamusa was saved from further slaughter.

The panic continued that night, even though the threat from the mob had faded. The family continued to worry about the welfare of Savitri and her family and other friends and relatives in the city.

But by the next day, August 16, people from the center of town, including Savitri, Brij Anand, and their infant, had come toward the police station and the street where the Sehgals lived. The family felt great relief in the midst of a growing sense of unease.

This day, the last day of Ramadan in 1947, was when the exact boundaries of the Partition were finally announced. The news angered various factions throughout the Punjab, Bengal, and elsewhere in the country. Many were not satisfied with the way their communities were being divided. Fear and violence skyrocketed everywhere in the region.

The immediate result was the start of what would quickly become a mass exodus in both directions. Hindus and Sikhs in the new Pakistan began moving toward the new India. Muslims in India headed to Pakistan.

In Lalamusa, the British commander, who had successfully defended the town, now declared the entire area from the police station to the railway station as a refugee camp. His soldiers erected a barbed-wire fence encompassing the area, about 250-by-500 meters, that stretched a little beyond the police station and included the temple, the Sehgal home, the gurdwara next door, a farmhouse, the primary school for girls, and the medical dispensary. People were told to stay inside the camp for their safety. A curfew was imposed. Soldiers were posted at the perimeters as guards.

The oppressive summer heat was almost unbearable. People bedded down in the open within the camp, sleeping wherever they could find space. The military patrolled the camp at night.

The Sehgal homestead was at the very center of the refugee camp, so the family felt lucky to be able to remain in their own house. They offered their hospitality graciously to everyone who came. Friends, relatives, and strangers sought shelter with them. Shahji and Shila extended the loyalty and generosity that Punjabis were known for. Each guest was an honored guest and treated warmly with all the courtesies.

One young woman in her late twenties, Mrs. Punjab Singh, ended up staying in the Sehgal home longer than expected and became a good friend amidst the turmoil. Her husband was a military officer posted in Meerut, near Delhi, in India. Military families didn't have much money, and she was anxiously awaiting the opportunity to leave the safety of the enclave. She planned to join her husband as soon as space in a refugee train heading for India was available. This contact would become a lifeline for the family before long.

The palpable panic and terror of the first couple of days, the violent mob and the creation of the camp, was replaced by constant busyness in the Sehgal household. Everyone in the family had to help out. Suri had ongoing chores to assist with dozens of people staying with them. There were cots to set up, meals to serve, and continual cleanup after so many guests. Wounded people were brought to the camp on a regular basis, and Suri helped in any way he could. Now under the protection of the military, a certain curious excitement accompanied the young boy's apprehension in this new routine.

Specially designated trains began to take Hindu and Sikh refugees to India and to bring Muslim refugees from India to Pakistan. Each refugee train reserved one compartment at the tail end for the army personnel. All refugee trains were escorted by the military, Muslim soldiers usually, if the train was coming from India; Hindu or Sikh soldiers, if the train was headed to India. However, this only meant eight to ten soldiers with guns, at most.

The trains passed through the Lalamusa station, but arrival and departure times were completely erratic because of the violence occurring throughout the region. The uncertainty added to the tension. The trains originated in Peshawar or Rawalpindi on their way to Lahore. After that, the trains headed east to India. All along the route the trains traveled through hostile territory; many were attacked before reaching Lalamusa.

Sometimes the trains arrived at their destinations full of dead and injured people. The casualties were taken off. The sweltering heat meant that dead bodies had to be left behind along the railroad tracks, where they rotted and were eaten by vultures. There was no other way. The trains were packed, with people literally sitting on top of each other and more sitting on the roofs of the train cars. Everyone was trying to flee the violence pressing down upon them. Trains in both directions were crammed with scared, tired people.

Those in the camp tried to leave when news reached them that a refugee train might stop in Lalamusa. There were frequent disappointments, however. As trains arrived, there was usually not enough room even after the dead bodies were taken off. People were desperate, even frantic. They piled into and on top of every train that stopped. Throughout the refugee camp, people pleaded for the chance to get on trains headed for India. Hundreds of people came to and went from the Sehgal home over the next few weeks as the horrors multiplied.

In areas without railway stations, departing Hindus and Sikhs formed caravans. Slowly and collectively, the convoys headed toward India from small villages throughout the Punjab. Many traveled on foot with as much as they could carry on their heads or in carts pulled by pairs of oxen. Belongings were piled high in every cart, with cattle driven alongside.

Stories came back to the camp about great carnage and violence on both sides of the new border between India and Pakistan. Bitterness was overflowing, and brutalities worsened. Muslims suffered at the hands of angry Sikhs and Hindus in India, just as Hindus and Sikhs were suffering at the hands of angry Muslims in Pakistan. The rioting continued for weeks as Hindus and Sikhs left West Punjab and Muslims left East Punjab. The violence stoked greater levels of vengeance and retaliation in each community. Hundreds of thousands of people on both sides of the border were butchered. This mass migration by so many had never occurred before in human history and remains unparalleled. The true extent would not really be known for many years after the fact. Somewhere between twelve and twenty million refugees were dislocated, and more than a million were killed.

Gandhiji, Nehru, and their Congress Party tried to stop the violence and promote peaceful coexistence, but the success of their efforts was minimal against this level of atrocity. Those at all points on the political spectrum were shocked and overwhelmed by the depths of the ghastly crisis.

The town of Lalamusa and the refugee camp were rapidly emptying. Anyone who could leave was doing so, despite the risk, by train, truck, bus, or bullock cart.

But in the face of all that was going on throughout the country, and the violent behavior surrounding them, Shahji remained reluctant to leave, even after several weeks. More than a few of the family's Muslim friends came to the house, urging Shahji to stay.

A staunch supporter of the Congress and a sincere believer in secular traditions and community spirit, Shahji held out hope that things still might improve. He worried that, as district president of the Congress Party, his departure might give the impression that he, too, was running away, or that there was no longer any reason for others to hope. He did not want to abandon the people in the camp. He would only agree to move if all the other people moved, too. In addition, the loss of their house in Lalamusa and all their other homes and properties in the surrounding villages, and in Guliana, would be substantial.

But as the trains came and went across the border, the sights of bloodied and dismembered victims stirred up the local people on each side to greater anger. Resentment and animosity festered as the situation deteriorated and violence worsened every day. Movement of any kind became very limited. Killings were reported on the outskirts of the refugee camp every day. Non-Muslims who dared to step outside the fenced area risked death, and those who left were not heard from again. Angry Muslims now sometimes lurked beyond the safety of the camp and carried daggers with poison on the curved tips. Because it was difficult to tell a Punjabi Hindu from a Punjabi Muslim (the two groups had no obvious physical differences), the men with daggers confronted those they encountered outside the refugee camp by drawing their blades and asking the men to recite a verse from the Koran. Others were forced to pull down their pants to show whether they were circumcised and therefore Muslim, and not in danger.

Known Muslim friends came and went, able to leave and return safely. Most visited for business reasons. The only Muslims who actually lived inside the refugee camp were the blacksmith and his son.

As long as the Hindu or Sikh military provided protection, there was no serious problem. But the military kept changing. There was always the chance that hostile Muslim soldiers would come.

Savitri's husband Brij Anand had a job in the military, and he felt confident that he could obtain some sort of vehicle to arrange transportation for the family. Savitri and the baby stayed with her parents when he left the camp to get help. He had no idea that it would be impossible for him to return for his wife and child.

The determining fear was obvious. Everyone was afraid for their own lives to one extent or another. But most important, families wanted to save their sisters and daughters from rape and abduction. Many girls and young women had disappeared throughout the region. Stories had come back that some Hindu and Sikh women jumped into wells, preferring suicide to that fate. Knowing it was very risky for his daughters to stay in their home, Shahji decided one day in September that the girls had to leave immediately, by whatever means possible.

Suri witnessed part of a conversation his father had with some men in the room near the stairs where the family usually played cards. A few Muslim men sat around the card table, discussing the possible purchase of some Sehgal properties. Shahji told them he was willing to sell the house and farm at any price, that he desperately needed cash for his children's travel. Suri recognized the sense of urgency and was alarmed by the look of desperation he saw on his father's face. Unfortunately, no deal occurred that day.

Late one evening only days later, Shahji learned from a Muslim friend who worked for the railway that a refugee train was expected to arrive from Rawalpindi early the next morning. It would be stopping only briefly in Lalamusa before heading toward India.

The elder Sehgal decided that four of his children would be on that train. Kedar, now twenty-five, had come home when the trouble reached Lahore. Shahji planned for Kedar to accompany his three sisters—Shakuntla, fifteen; Padma, almost fourteen; and Santosh, eleven—and keep them safe on the train ride to India.

Savitri and the baby would stay, waiting for her husband's return. The two youngest girls, Parsanta, age seven, and Sanjogta, almost three, would also stay behind with Suri and their parents. The departing children knew nothing about it until the September morning when their train was supposed to arrive.

Shahji and Shila woke the girls early and told them to put together a few things, "not much." They were told, "The train is coming. Get ready now."

Suri was awakened by his father and told he was to accompany his brother and three sisters to the train station to help with their baggage. Shahji and Suri would return home after the four were safely on the train.

There was no concrete plan for what would happen once the siblings reached India, nor could there be, under the circumstances. None of the family's relatives were living in India at that time, except one uncle. Shila's brother Gurdit Singh, a police inspector, was now working "somewhere near Delhi." But they had no address for him. Their close family friend, Amar Nath, lived in Palampur. Otherwise, the whole extended family was spread out from Peshawar in the North to Lahore in the South, and none further east than Lahore. The only address they had was Mrs. Punjab Singh's, the young woman the Sehgals had taken in and helped when the refugee camp was first established several weeks before. Kedar and his sisters had only a few rupees between them and no idea if or how they would find their parents again once they reached India—if they reached India safely.

Suri and his father went to the station with Kedar and the girls. Suri helped carry the suitcases they'd quickly stuffed their belongings into. But when the train arrived, it was already crowded. There was no way for Kedar and his three sisters to fit in one compartment. The baggage they were carrying could not go with them; there was simply no room. They could each only bring along a small bag of clothes. Everyone had to move fast. Shakuntla and Padma managed to squeeze into one compartment, and Kedar ran into another.

Eleven-year-old Santosh found a spot in a rear train car. But someone needed to accompany her. Suri was enlisted at the last minute as Shahji quickly pushed the boy into the compartment behind his little sister and, without warning, the train started moving.

Shahji hurriedly said good-bye to his children, literally running from one compartment to the other as his dear ones left for an uncertain location and a perilous future.

Suri had, of course, not left the house that morning with any expectation of leaving on a train. He had nothing at all with him when he was tossed like a football onto the train car by his desperate father. Suri had only the clothes he was wearing: a pair of knickers (shorts), a short-sleeved shirt, and flimsy sandals.

Instantly accepting his assigned duty, despite his own panic, Suri struggled through the crowded train car to reach his little sister. When Santosh saw her brother, a degree of relief could be seen along with the terror in her eyes.

Suddenly dislocated from the intimate domain he had enjoyed until age thirteen, Suri would now have to rely solely on his own inner sense and the values embedded in him by his family and his childhood in the Punjab.

CHAPTER 4

Blood on the Tracks

The crowded train made its first stop in Gujrat. Though Gujrat was only about twenty-one kilometers from Lalamusa, it had taken half the day to get there. Unbelievably, even more people climbed aboard the train. Suri could not imagine how they fit because people were jammed in together so tightly already that no one could sit down. Every spot was taken on the crowded roof of the train as well.

The train began to move again very slowly. The steam engines had to take on water and coal wherever they stopped. The next station was Gujranwala, about fifty-one kilometers past Gujrat. After refueling at this stop, the train didn't move. No one on board knew why. Standing for such a long time was exhausting. The stifling heat and humidity of the monsoon season, though still no rain, added to the ominous mood. Fear intensified. Everyone on the train had heard horror stories about trains that were made to stop, and were then attacked. A stopped train gave angry locals the opportunity to assemble in groups to attack, kill and loot, taking whatever cash and belongings people had. Every passenger was on high alert, numb with anxiety. There was no conversation, only fear.

There was nothing to eat on the train, and nothing could be purchased at the train station. The water in the train car washrooms was not drinkable. The only drinking water available was on the train station platform, but the passengers feared that the water fountain was poisoned. No one dared to venture into town where shops might have something to eat or drink.

The soldiers ordered everyone to remain inside the train throughout the evening and all night. The discomfort and fear created an almost unbearable tension. The desperate refugees could hear shouting in the distance, the same shouts Suri heard back in August from the screaming mob in Lalamusa in front of his house. All evening, voices echoed from the nearby village, shouting, *"Allahu Akbar! Allahu Akbar!"*

To the refugees on the train, that phrase signified danger—a rallying cry used by Muslims to gather the faithful—as if to say, "Come! Come, the infidels are here!"

The train remained in Gujranwala the entire night. At some point the shouting stopped, and no attack came. Despite hunger, thirst, and fear, everyone was relieved when morning came, including the soldiers. But the refugees were getting desperate without food, water, safe shelter, or rest. Without explanation, the train stayed stuck in Gujranwala station for another day and night.

Everything remained quiet outside the train, but an attack was again anticipated at any moment. People on the train remained still and silent, enveloped in dread. Everyone knew they could be killed at any moment; but there was nowhere to run, no way out of this terror-filled train trip that, hope against hope, might end somewhere safe. At this point, the brothers and sisters lost all concept of time.

The ongoing fear and deprivation made the fleeing refugees feel like they were living inside an endless nightmare. To protect them from the bullets, the people closed the windows and blocked them with whatever luggage they had. The overflowing latrines on the train filled the air with stench. People were suffocating.

The siblings found each other outside from time to time on the train platform. At one point, Suri's sister Padma declared, "I am going to die. I am going to die, really!" just before she fainted.

Kedar rushed off the platform to a pond nearby and brought back some water for his little sister. As she gratefully sipped the water, Padma said, mostly to herself, "I am alive."

The dirty water in the nearby pond was the only water anyone drank. Obtaining a scrap of food was at the mercy of the soldiers riding with the train. The military had limited supplies for themselves. The soldiers and guards tried to be kind, giving what they could spare to the civilians, mostly dry biscuits or an occasional sip from individual canteens they carried for their own use. The most anyone ate was a single spoonful of rice.

Suri was grateful to be in the compartment next to where the Hindu military escort rode, the last car before the caboose. Getting off the train was dangerous, but occasionally necessary—at least to get some pond water. But even when Suri walked outside the train with one of the soldiers, he didn't really feel safe.

Trains full of refugees coming from the other direction were crammed with Muslims who had witnessed horrific violence in India. Every passing train evoked fear of retaliation. While their train was parked at Gujranwala, Suri heard the familiar cries erupt as the trains passed: "Death to infidels! *Allahu Akbar!*" Luckily none stopped.

Finally Suri's train moved on, very slowly. The next stop was Lahore, where the train stayed for only a few hours before moving on again. Arriving in Lahore on this trip was not associated in any way with the carefree fun Suri had experienced playing with his cousins there over the years.

Though the Sehgal family did not learn the details until much later, Suri's Sikh aunts and their families had left Lahore months earlier—one went to Delhi and the other to Simla, the seat of the new government of East Punjab, where her husband continued to work with the government. They had originally thought, as others had, that the move was temporary, just for a few months. They fully expected to be able to return after things settled down, but the family never returned to their homes in Lahore.

After Suri's train left Lahore, it was expected to continue on to Amritsar, the most important border town inside India. But inexplicably, the train headed south toward Ferozpur, a smaller town on the India side of the border, after making a brief stop in Kasur, a Pakistan border town. Passengers on all refugee trains were basically dumped in whichever town was closest to the border inside the destination country. From that spot on, they had to fend for themselves. That is what happened to Suri and his siblings after many days on the crowded train.

Only a few days earlier, empty trains spattered with blood had arrived at the Ferozpur station. Armed mobs had stopped the trains in

Kasur between Lahore and Ferozpur and slaughtered almost everyone on board. The same thing happened to trains between Lahore and Amritsar as well. Thankfully, this train did not meet such a fate. The Sehgal family saw no violence upon their arrival in Ferozpur. Instead, the train was met by Hindu and Sikh volunteers offering food, water, and relief. Anyone who wanted to move into the refugee camps could do so right away.

Suri and his brother and sisters were happy just to be alive and on the other side of the border. But they were shocked by the enormous number of dead men, women, and children they saw along the train tracks. Putrefying dead bodies lay on both sides of the railway, and the stench filled the air. They had not seen bodies along the tracks from Lalamusa to Lahore. The evidence of savagery that they now witnessed from Kasur onward was mind-boggling. The family learned that within the district of Ferozpur, Hindus and Sikhs had hacked to death hundreds of Muslims in retaliation for equal horrors inflicted on the other side of the border at Kasur.

Suri realized how lucky they had been on their terrifying train ride. Though his sisters had seen four or five people on their train die of heart attacks in the heat, stress, and conditions on the trains, four or five in a train of hundreds didn't seem so awful compared to what they were seeing now.

After gathering together and talking for a few moments, the brothers and sisters decided to continue on. Their goal was to eventually reach Meerut, and the home of the lady their family had helped in Lalamusa: Mrs. Punjab Singh.

No train was going directly to Meerut, but there were trains going in that general direction. The idea was to catch whatever train went south. For five people to board the train together was still not easy. The Sehgal siblings had to push their way into compartments to find space on the crowded trains. They had to change trains two or three times to travel 424 kilometers to Meerut, which was 70 kilometers northeast of Delhi. They first rode a cargo train, scattered in different compartments. Suri found a spot in the caboose reserved for the railway guards. A train guard there did not permit him to sit inside, but let him sit on the steps outside with his legs dangling. Unfortunately, Suri lost a sandal on that ride. Still wearing the same pair of knickers and short-sleeved shirt he left home in days earlier, the boy was now barefoot as well.

After a seemingly endless and uncertain stop-and-go journey past more rotting bodies on each side of the train tracks, the brothers and sisters finally arrived in Meerut. They made their way to Mrs. Punjab Singh's house, where they were warmly welcomed. She offered what she could, expecting them to stay only a couple of days and then move on to Delhi to locate their uncle, which was their plan.

But the monsoon, which was so late in coming, finally arrived. The rains thundered down on the parched land with ferocity. It rained and rained for days. Bridges over streams and rivers were washed away, including the bridge over the Yamuna, the river separating Meerut from Delhi.

With no way to reach Delhi, the siblings were stranded at Mrs. Punjab Singh's house. Now the helpful hostess became anxious and eager for her guests to leave. She was not prepared to feed five additional mouths for more than a few days.

Soon Kedar and Suri each took on new responsibilities. Kedar would immediately return to Amritsar, where most of the refugee trains were arriving, to learn any clues about the whereabouts of the rest of the family. Their train had been an exception, ending up in Ferozpur. Kedar hoped that, by now, Shahji and Shila and his other three sisters had managed to escape from the refugee camp in Lalamusa and were now safe in Amritsar.

Suri's task was to go to Delhi and locate their uncle, Gurdit Singh. But the only information Suri had was that his uncle was a police inspector somewhere near Delhi—no specific station, nothing else. Suri had to first figure out how to get to Delhi, without any money, then figure out how to find his uncle.

Suri's sisters had nowhere else to go. They had to stay with Mrs. Punjab Singh, who by this time was not at all pleased with the situation.

The surprising last-minute arrival of a friendly face gave Suri a welcome traveling companion on his quest. Gurbaksh Singh, the eighteen-year-old family friend with whom Suri had happily played sweep in Lalamusa a few weeks earlier, showed up on Mrs. Punjab Singh's doorstep. Gurbaksh had taken a refugee train earlier than Suri's train and ended up in Delhi. He had heard from a refugee family in Delhi that the Sehgal brothers and sisters had arrived in Meerut.

Gurbaksh, profoundly indebted to the Sehgal family for their help with his education, now was keen to render whatever assistance he could in their current difficult circumstances. Deeply engrained with the Punjabi cultural imperative to find ways to return any favor someone has done for you, Gurbaksh was happy to have this opportunity. He suggested that Suri and he go together to Delhi, and he would be pleased to help Suri find his uncle. Suri was delighted to see his friend and to have a traveling companion.

Without food, money, or shoes, Suri set out for the capital with Gurbaksh. The boys rode a free bus to the bank of the Yamuna, on the eastern edge of the capital. The bridge was still out, but by this time the river had receded quite a bit and the teenagers managed to wade across. On the other side, they took another bus into the city of Delhi. The capital had a population of about one million people at the time, with additional refugees streaming in every day. The city was in constant turmoil.

In Delhi, Gurbaksh used what little money he had to buy Suri a pair of shoes. The boys then headed for Gurdwara Sis Ganj, a well-known Sikh temple and historical holy shrine located within the Chandni Chowk marketplace that was just down the street from the railway station. Gurdwara Sis Ganj served as a place of prayer and worship as well as a community kitchen, serving meals to the homeless. Anyone was welcome.

Delhi was like a foreign country to Suri. He felt suddenly overwhelmed by the seeming impossibility of his mission. To find one person in the midst of all these people, without a clue to his whereabouts, was a daunting assignment. Separated from his brother and sisters, the only relatives he knew for sure were still alive, Suri was miserable, but he knew he had to locate his uncle, somehow, in the midst of this new and scary environment. He was very grateful for Gurbaksh's company.

Suri's sisters were similarly isolated in Meerut with their unhappy hostess. They worried about their young brother. They had no idea where he was or if he was safe, only that he had gone to Delhi with Gurbaksh Singh.

Suri had arrived in Delhi at the start of the Dussehra festival, which fell on October 11 that year, and he was present when the new prime minister of India, Jawaharlal Nehru, arrived at the celebration looking larger than life on a white horse. This was the first time Suri saw Nehru in person after all the years of hearing about him from Shahji and seeing his photo in the newspaper. Suri did not speak with Nehru, but he watched as Nehru talked with others in the huge Ramlila Maidan (Ground), a popular site for large gatherings, where the demon king Ravana's effigy was about to be set ablaze.

Since Suri and Gurbaksh had nowhere to stay, they slept in the railway station, in trains that were parked in the yard. They would spot a certain passenger train that was not moving, and sleep in one of the compartments. On two occasions, the train started moving while they were asleep, and they woke up in other towns! The shudder and lurch of the train stopping in the town of Mathura woke the exhausted boys the first time it happened. They jumped off the train to find hawkers of tea and food on the train platform. Luckily, travel without a ticket was not a problem during this tumultuous time since so many displaced people had little ability to pay. Suri and Gurbaksh hopped the next train back to Delhi.

The boys decided to find an empty house to sleep in instead of train cars. Fleeing Muslims had abandoned their homes in India when they escaped to Pakistan, just as many Sikhs and Hindus had fled their homes in Pakistan. Suri and Gurbaksh spent an afternoon searching for such a house. However, there was a catch. In order to claim a house for more than one night, they would need to buy a lock. Neither boy had money to buy a lock. As they walked along the street in the once-Muslim neighborhood of Paharganj, the boys saw two men walking ahead of them. They were dressed in the garb worn by the most conservative Hindus. However, someone on the street recognized the two men as Muslims and began shouting and pointing at them.

Suri thought the Muslim men had perhaps wanted to return to their vacated homes to retrieve items left behind and were wearing the disguise for protection. The men had not expected to be recognized.

A nearby policeman, a Sikh, took out his gun and shot into the air. That was his only effort to disperse the Hindus and Sikhs rushing together and circling the Muslim men. The policemen did nothing to stop the people who grabbed the two men and dragged them toward the now-empty local mosque.

Suri and Gurbaksh followed as the men were pulled into the courtyard of the mosque. The boys were horrified when four or five men suddenly picked up large rocks and smashed the Muslims' heads. Then, methodically, step by step, the men hacked off the Muslim men's limbs with swords and knives, turning both men into piles of bleeding body parts.

Standing only a few meters away in a crowd of forty or so people who were watching and not objecting in any way to the murders they witnessed before their eyes, Suri and Gurbaksh were aghast and sickened.

The thought screamed in Suri's mind, *How could anyone do this?*

Suri had seen many injured people in the camp in Lalamusa, many with severed limbs, but this was the first time he actually witnessed up close the appalling actions of murderous angry men who could do such things. Suri and his friend stood frozen—revolted, terrified, and help-less, with tears streaming down their faces.

Turning, finally, the boys walked away. They returned to the rail-road yard, abandoning their search for an empty house. The nightmare of that day, along with the gruesome images and his profound disgust at such violence, would stay with Suri for the rest of his life.

The boys needed money, and Gurbaksh thought of a way to try to get a job. He knew that displaced former government employees from the Punjab could automatically report for certain types of jobs in Delhi. Gurbaksh went to the railroad people and told them he had worked as a train guard with the railway before Partition.

He had never actually been a guard. He didn't even know what a guard did. But he had found somebody's official insignia and took it with him to the interview for a job that would pay sixty rupees per month (about twelve dollars).

When he was asked what proof he had that he'd held such a job, he said, "Here it is! This is the only thing left from my uniform, this insignia."

"Well," he was told, "I don't know anything about that insignia . . ."

But Gurbaksh was offered a job as a clerk. He would get a paycheck, but he had to wait until the end of the month before he received the first check.

Now that Gurbaksh was a railway employee, he was provided with a modest accommodation, a tiny one-room flat with a little kitchen, but no bathroom. The boys still had to use a community bathroom at the end of the street. Fortunately, the weather was warm enough that Suri could sleep outside on a cot in the courtyard.

Suri had been wearing the same clothes since leaving Lalamusa. The first thing Gurbaksh did after receiving his first month's pay was to buy Suri a new shirt, knickers, and underwear.

Mahatma Gandhi was staying in Delhi at Birla House, the home of his close associate and admirer, Ghanshyam Das Birla, a wealthy and influential businessman. Gandhiji occupied a room on the ground floor. In the evenings, for an hour or so, anyone could join Gandhiji on the prayer grounds adjoining the house where he prayed and preached peace.

Suri rode the free bus there, sometimes with Gurbaksh. In the midst of such carnage and uncertainty in his young life, Suri found it comforting to hear Gandhiji preach the same message of unity and harmony that Suri had heard all his life from his father.

Gandhiji said prayers that included scripture readings and prayer chants from multiple religions, with a few words that Suri recognized from other sources, including the Bible and the Koran. Suri considered Gandhiji a highly respected saint, as most Indians did, and felt both comforted and honored to be in his presence. In the company of Gandhiji, the lonely boy renewed his commitment to his task. As was his nature since childhood, which seemed so long ago at this point, Suri would keep trying different ways to solve his current dilemma. He knew that giving up was not the answer to survival. He would continue to be resourceful in the hope that he would someday see his family again.

While Gurbaksh was at work, Suri continued his search for his uncle and also for sources of food. He met a friendly young man who was working as a volunteer helping refugees. Recognizing Suri's circumstances, the fellow offered to take Suri to a friend's house where there was "very good food." Suri felt comfortable enough with the guy to go with him to the home of a well-to-do Hindu couple who had no children. They gave Suri the best meal he'd eaten in a very long time: tasty food, delicious tea, and wonderful sweets. The people seemed intent on making the lost refugee comfortable in their lovely home. After an hour or more there, though he did not know why, Suri was feeling a little uneasy, so he thanked the kind people and excused himself, saying, "Now I have to go."

When the couple told him they'd like him to stay, he said he would come back later and he made a hasty retreat. Suri had lived in the streets of Delhi long enough by then to recognize yet another type of solicitation. There were people all along the streets who tried to get young refugee boys to do lousy jobs with no pay, or any number of other things Suri had no time for. He had no interest in being taken in by this childless couple, regardless of their good intentions.

The young man who had taken Suri to meet the couple followed him back to the train station in Delhi. He asked Suri, "Why didn't you stay? Those people had such a nice house . . . you could have stayed with them! They are very nice people with plenty of food and space! They even offered to send you to school! They wanted to adopt you."

But Suri already had a family, and he had a job to do. His sisters were still stranded. He had to find his uncle.

CHAPTER 5

Miracles

Suri and Gurbaksh continued to search vigorously for Gurdit Singh, going into police stations and asking questions. Finally, at one station, the boys learned there was indeed a Gurdit Singh, inspector of police, living in Ballabgarh, quite a few kilometers southeast of Delhi, just past Faridabad. Full of hope, the teenagers went to Ballabgarh and walked into the police station. There he was! He was real—his uncle, Gurdit Singh! Suri's uncle looked quite surprised. He left work immediately to take Suri back to his house.

To the thirteen-year-old who had just survived so much hardship and uncertainty, being reunited with family was a huge relief! Suri was elated.

Meanwhile, Suri's sisters, Shakuntla, Padma, and Santosh, though grateful they were safe, were not at all comfortable staying with Mrs. Punjab Singh. They were made to feel increasingly unwelcome. Their hostess made no secret of her displeasure at being saddled with the three young girls. After their terrifying trip from Lalamusa, the violent images and the deprivation, and being apart from everyone else in their family, all they could do was cling to one another and cry, further annoying the already frustrated Mrs. Punjab Singh.

After a few weeks, Mrs. Punjab Singh reached the breaking point. She would not allow the girls to stay any longer. She wanted them to leave immediately. The sisters, still terribly sad and frightened, agreed that they had to leave Meerut, but they had no idea where to go.

Mrs. Singh told them bluntly, "You have an uncle. Go to Delhi. You'll find him!"

Mrs. Punjab Singh arranged the transportation with her husband, an officer in the army. The girls were driven in a military jeep to Delhi and dropped off at the Gurdwara Sis Ganj near the railway station—the same gurdwara where Suri and Gurbaksh found refuge and sustenance when they first arrived in Delhi several weeks earlier. The sisters could get free food there, but nothing else in terms of assistance or lodging.

The abandoned young girls sat frightened in the gurdwara, thinking, "What do we do now?" when, just at that moment, Shila's oldest brother, Doctor Kartar Singh, happened to walk into the gurdwara to pray! Profoundly delighted and relieved to see a familiar face, the girls were also amazed at the unbelievable serendipity of their uncle's arrival at that exact time! They considered this a true miracle! The doctor had come to Delhi from Nowshera in a military convoy from Pakistan. He was very pleased to find that his young nieces were safe, but he knew nothing about where Suri or anyone else in the family was.

A follow-up miracle involving the Sikh prayer practice in this particular holy shrine answered this question within moments. Gurbaksh Singh had taken his regular break from his job as a clerk at the nearby railroad station at just that time to pray at the very same gurdwara. The sisters immediately recognized their old neighbor, just as he recognized them. Gurbaksh excitedly told them that Suri had just found, and was now staying with, the doctor's own brother, Gurdit Singh!

For this amazing combination of wonders to occur within the Gurdwara Sis Ganj, a holy place that had been known as an ultimate symbol of unity between their Sikh and Hindu ancestors since the 1600s, was a fitting brush with destiny. The pure miracle of the blessing that brought them all to this shrine at the same critical moment was never forgotten by the Sehgal family.

Kartar Singh was able to contact his brother Gurdit Singh in the police station, using the military convoy phone. Gurdit Singh showed up in a police jeep and took all of them to Ballabgarh. The sisters rejoiced. They were so happy to see that Suri was safe, and that he now had shoes! By this time Gurdit Singh had provided Suri with fresh clothes. Now he would buy clothes for each of his nieces as well.

Suri and his sisters were relieved to be reunited and with their uncles, but it was time to figure out what step to take next to find the rest of the family. There had been no word from Kedar. He had not returned to Meerut nor come to Delhi. Suri needed to go to Amritsar to find out what happened to his brother and try to locate his parents. Kartar Singh offered to let his nephew ride with him on a military convoy to his new post in Amritsar.

Shakuntla, Padma, and Santosh stayed with their uncle Gurdit Singh in Ballabgarh, where they were welcome, and where they were eventually reunited with their maternal grandparents as well. Rawel Singh and Har Kaur had left Arah before the Partition for an extended visit with their son Gurdit Singh. By this time, the elder Wadhawans were both somewhat infirm. Sadly, the doctor's arthritis was much worse, and he was in pain a lot of the time. He was still staying with Gurdit Singh, but Har Kaur had gone to visit another son, Harbans Singh. The girls were overjoyed to finally know for sure that more of their family members were still alive.

When Suri arrived at the refugee camp in Amritsar, he found out why his brother Kedar had not come back to Meerut or Delhi. Their parents, it turned out, were still in Pakistan.

Kedar had connected with a friend in Amritsar who was working with the government in the weights and measures department. While staying with his friend, Kedar was trying to make arrangements for his family to be able to stay in one of the vacant houses in town once they arrived, but he had not seen his parents yet. He had been coming to the refugee camp every day, anxiously hoping for their arrival. He looked for them at the railway station, checking each train, and asking questions of anyone from Lalamusa.

The refugee camp was a large open area, and hundreds of people were staying there. Suri and Kedar split up to cover more ground, searching for their parents among the continuing arrivals from Pakistan. They spoke with dozens of people and asked lots of questions. Everyone they met had a terrible story to tell. Kedar kept watch at the train station, but somehow they missed each other in late November when the family finally arrived. Shahji and Shila and the girls ended up in the refugee camp, which Suri was scouring on his daily vigils, looking for familiar faces, and shivering in the cold as he walked up and down the rows of families. Folks were wandering around, and many were sitting on small floor mats, creating a patchwork of people and colored cloth that completely filled the huge open courtyard.

Though Suri had no warm clothes, he was fully warmed when he saw his dear mother and his little sisters, seven-year-old Parsanta and three-year-old Sanjogta, sitting together under the veranda of the square. Rushing over to them, Suri could see right away that his mother was not well. Her face looked drawn, and her eyes looked weary. But Shila was elated to see her young son, and the two exchanged tender hugs of relief and excitement. Shila had been ill since before leaving Lalamusa. The stress and terror had taken a toll. Shila shed tears of happiness as she blessed Suri. The little girls looked well and were full of smiles seeing their big brother.

Suri was told that Shahji was off trying to reconnect with other relatives, former colleagues, and friends, and figuring out what the family's options were in these new circumstances. Before long, he returned, just as Kedar found them. Shila cried more, hugging her oldest son. Kedar and Suri each touched their father's feet in respect and affection. Shahji was very pleased to find his sons safe and well and to know that the three girls were safe with their uncle. Both parents offered blessings to their happy sons, touching the tops of their heads.

Shahji, Shila, and the girls had already been at the camp for a couple days. Brij Anand had found Savitri and their baby; they were now staying with his relatives, who had arrived in Amritsar a few weeks earlier.

In the initial commotion of their family reunion, sharing who was where, the family was immersed in feeling quite blessed that everyone was alive, especially in light of how terrible things had gone for so many since the morning many weeks before when a refugee train took half of the Sehgal family away from their home in Lalamusa. Amazingly, Suri's entire immediate family was alive and safe.

When Suri and Kedar asked what their parents and sisters had endured back home, Shila began to sob. Shahji gestured for Kedar to take a walk with him so they could speak privately about the terrible details of what had occurred.

While Shahji and Kedar were gone, Suri comforted his mother. Suri and Shila had always been able to speak easily with one another. Shila was a wonderful listener who could find common ground in any conflict and peaceable solutions to any problem Suri told her about.

Shila asked him to sit next to her and tell her more about his other sisters and the rest of her family. She wanted to know every detail about her daughters, brothers, parents, everyone! Though she kept weeping, Shila was overwhelmed with relief as Suri related stories about their time apart. Shila and Suri talked together with joy and tears until Shahji and Kedar returned from their walk.

Kedar later filled in Suri with much of what their father had said about the family's final exit from Pakistan. The Sehgals had been among the last to leave their house. Most of the people in the Lalamusa refugee camp had eventually left. The Hindu army soldiers had been protecting them for weeks, but an abrupt change of regiment meant that all Hindu and Sikh soldiers migrated to India, and Muslim soldiers took over. The military that was supposed to protect the camp were now all Muslim. These Muslims were very different people than the Sehgal family's Muslim friends and neighbors. The new regiment was a tribal group from Baluchistan, a Muslim region in the southwest known for its law-lessness. Baluchis were known throughout India at the time as ruthless plunderers and marauders. These soldiers fit that reputation; they were hostile and violent.

Since the Sehgal home was large, the soldiers assumed these home-owners must have a great deal of cash, jewelry, and gold. The Baluchi soldiers entered the house one night, shouting harshly and wielding their guns, demanding, "Give us all your money!"

They screamed that they were going to kill everyone in the family. They immediately threatened to take three-year-old Sanjogta, shouting, "We will cut this little girl into pieces!"

Shahji and Shila were terrified that the soldiers would abduct or harm Sanjogta, Parsanta, or Savitri's baby, who was asleep in Shila's arms. They were scared also for Savitri, who was twenty-one and attractive, and a likely target of the soldiers' exploitation and abuse. When the soldiers came, she had crawled in between the stacks of standing cots under the veranda to hide. Now that most visitors had left, there were close to a hundred empty cots stored there. Since there was no electricity in the house and it was pitch dark, she successfully escaped attention from the Baluchis. They never found her, even though the soldiers roamed the house all night long, searching for valuables.

In truth, the whole family narrowly escaped death. Shila later told Suri how the soldiers had brutally beaten his father again and again throughout that dread-filled night. The Baluchis were trying to make sure they found every single thing of value in the house. Each time they found something else, they beat Shahji further to make him tell them where to look for more. They took all the jewelry, gold, and cash, leaving nothing but a few pieces of clothing.

The next day, several Muslim friends who lived on the same street helped Shahji and his family catch a noon train. Shila was overcome; the relentless terror had left her ill with a fever. Feeling weak, she could not even walk, so Muslim friends kindly arranged a tonga to take her to the station and gave the family some food to take on the train.

Shahji, Shila, Savitri, and the three children all managed to get on that noon train, and there were no attacks, though the train stopped many times along the way. Finally, they arrived in Amritsar, not realizing Suri and Kedar were waiting there for them.

In the refugee camp, there was no place to sleep. The food was free, but basic and limited—a bit of bread and *dal* (lentil soup). Colder weather was coming, and the family needed blankets. One of Kedar's friends, who was responsible for distributing blankets to the refugees, brought blankets and said, "Shahji, these will help your family stay warm."

But Shahji, who had no money to offer in return, refused to accept the blankets. He replied, "Don't call me Shahji anymore. I am no longer rich. You can call me Faqir, because I am now a poor man. I have lost my land, my house, and all the cash and jewelry I had. Faqir is my true name."

Shahji's main focus now was the immediate need for food and shelter for his family. He had to find a way to get them out of the Amritsar refugee camp and to start life in a new country. The next steps were clear: everyone in the family had to do what they could to find ways to make money.

Kedar went in search of a job. Suri was given the task, one he already had a little experience with, to locate an empty house large enough for his family.

Though many houses in Amritsar had been vacated by Muslim refugees, the competition for houses was heavy. Many Hindu and Sikh refugees were all seeking homes for their families at the same time. Suri searched the surrounding neighborhoods and finally found a tiny place the next day, not a particularly nice place, but it would provide much-needed refuge in the cold weather.

Now that his family had shelter, Shahji gave Suri a key task, acquiring basic supplies. Immediately after the Partition, staples such as flour, sugar, oil, and coal were in short supply. Suri had to go to a government office about three kilometers away in the district court compound to obtain a permit to buy those items. This was no easy task when thousands of other people were trying to do the same thing. The boy learned quickly that he had to be persistent and never to take no for an answer.

Large crowds of people waited for hours to get close to the small window where permits were issued. Suri recognized right away that he had to find a way to get to that window as quickly as possible. He wiggled his way through the crowd, and once he made it to the window, he kept asking for what he needed, saying, "Please, please, please," until his cajoling had the desired effect. His effort required tremendous determination, and the young boy was consistently up to the task. Suri somehow always found a way to get the required permit and the supplies.

Having taken on greater responsibility within his family, Suri seemed to have obtained a more adult status than usual for a thirteen-year-old boy. Recognizing Suri's determination and dependability, his father relied on his young son's skills in acquiring many basic necessities, while his mother relied on him completely as her confidant. Though the family was now safe, Shila was still not well. She was clearly shaken by the events of the past few months. Each day the family lived in the little house, Shila spent hours telling Suri the same horrible stories over and over: how Shahji was beaten, how Suri's sisters were almost killed, the atrocities she witnessed, the fear she had felt. She provided more details than Shahji had first shared with Kedar. Shila did not want to further frighten the girls by repeating the stories to any of them. Suri was her main sounding board and supportive listener and confidant, just as she had always been for him.

Shahji was successful in reconnecting with associates and friends in business and from the Congress party. His best friend, Amar Nath, still feeling obligated to Shahji due to the failed cigarette business, came to lend a helping hand. Once again, Shahji was blessed by friendship and the benefits of many returned kindnesses from those he had helped so generously in the past. With the assistance of well-connected friends, he was soon on his way to becoming somewhat established in Amritsar.

Government programs had been created to assist refugees and provide some compensation to families who'd lost everything in their desperate escapes. Those who could prove that they had owned a house or land in Pakistan could file a claim of sorts. In most cases there was no way to actually prove ownership of anything left behind. Shahji's only recourse was to swear under oath that he had owned homes and land, with only a hope that he would eventually receive some fraction of the amount the family lost.

The refugee program allotted each family a home to live in, so the Sehgals moved in January into the second floor of a slightly larger house than the one Suri had found; another family lived on the ground floor. By this time, Suri's other three sisters, Shakuntla, Padma, and Santosh, had arrived from their uncle's home in Ballabgarh and were reunited with the family in Amritsar. Shila was full of joy seeing the rest of her daughters again. Though her former robust health never fully returned, Shila's will remained strong. She kept the home together for her family as Shahji continued to collaborate with others to rebuild some security for his family and assist in helping his new community.

Shahji and Shila now wanted their children in school. Shakuntla had already finished eighth grade and could not go further without learning English. Padma, Santosh, and Parsanta were enrolled in schools for girls. Suri went to the local government high school and enrolled in ninth grade. He began to apply himself to his studies in earnest at this point. Because there was no electricity in the house at first, Suri spent many evenings studying under a streetlight, in the park, or wherever he could find available light. He worked hard.

Padma, who had turned fourteen by then, was in the same grade as Suri, but in a different school. Less than a year older than Suri, Padma was a brilliant student and envious of her brother because, as a boy, he could go outside in the evenings whereas the girls were stuck at home. The only quiet place Padma found to study was up on the flat roof of the house. However, Sheila and Shahji were strongly supportive of their daughters' educations. Even though girls couldn't safely go outside at night, their parents brought them tea and snacks while they were studying, even when they were up on the roof.

The violence that had begun to calm down on both sides of the Partition border by that time threatened to reignite one evening at the end of January 1948. The people of India and the rest of the world were shaken by news of the assassination of Mahatma Gandhi. Beloved Gandhiji was dead. Only two days before, the Father of the Nation had said, "If I'm to die by the bullet of a mad man, I must do so smiling. God must be in my heart and on my lips. And if anything happens, you are not to shed a single tear."[13]

In the moments between hearing the news of Gandhiji's death and then learning who had killed him, there seemed to be collective breath holding in India. If he had been killed by a Muslim, it would have no doubt meant more slaughter. However, Gandhiji was murdered by a Hindu, one of his own who apparently did not approve of the leader's tolerance of Muslims. Gandhiji died a martyr, still trying to bring together the diverse religious factions in India.

Shahji learned the news when he took his early morning walk to pick up a newspaper the following day. But Suri was already at school, so he learned the awful news from his school's headmaster, who called together all the students into an open area outside. When the headmaster announced Gandhiji's death in an emotional speech, every student burst into tears. Suri saw people were crying openly in the streets as he walked home.

Since school was a half day on Saturdays, Suri arrived home early and found men gathered there. With the news of Gandhiji's death,

---

13. "Gandhiji Shot Dead." *The Hindu*. reprinted January 31, 2013. http://www.thehindu.com/opinion/op-ed/gandhiji-shot-dead-the-hindu-january-31-1948/article4358055.ece.

some of Shahji's community and work associates, as well as two of Suri's uncles, had rushed to the Sehgals' house in Amritsar. The men shared their profound grief at the death of their much-loved hero. Suri saw how deeply saddened everyone was; the anguish was written over all the men's faces.

Suri heard his physician uncle Kartar Singh say, "Thank God the murderer was a Hindu. Had Gandhiji been killed by a Muslim, his death would have inflamed Hindus and Sikhs to even greater violence against Muslims."

The men agreed sorrowfully.

The entire world wept at the loss of Mahatma Gandhi. Everyone in India was in shock at the death of their saint and the man who had fought for peace to his dying breath. Yet in some ways the violent murder of this seventy-eight-year-old man may have played a part in calming the blood lust that had embraced the country for so long.

CHAPTER 6

Doors Open

By March 1948 the Sehgals were reassigned a bigger home in Amritsar for their large family and provided with some land in the Ambala District as part of their allotment. However, the people living in the newly assigned home were close family friends, a newspaper publisher and his family, who would be required to relocate if the Sehgals moved in. Shahji told his friends to stay as long as needed. The family living on the ground floor in their current home moved away at that point, which allowed the Sehgals to have enough room where they were for the time being.

While Suri and his sisters focused on school, Shahji had created a new base of operations and now had a source of income. With the help of his connections in the commodities supply department, he had acquired the necessary license to establish a food shop in a bazaar on the outskirts of Amritsar, among other retail shops. Though he had previously worked in wholesale, Shahji now worked in retail, selling flour and sugar. People needed a ration card to purchase these items because of the severe shortages, which usually meant they stood in long lines to buy these necessities.

Shahji's permit to sell flour and sugar was obtained jointly with another applicant who became his business partner. This arrangement was an advantage to Shahji because his new partner was well suited to do the required heavy lifting of the flour and sugar gunny (jute) bags that were delivered to the shop. He could more easily do the careful work of scooping the flour and sugar from the bags and weighing the individual allotted amounts for each sale. Shahji's strengths were in handling the finances, business negotiations, and correspondence. He worked long hours, but regardless of how late he returned home, Shila had a meal ready for him.

Shahji was offered the opportunity to have a liquor license, which he knew would be a lucrative enterprise, but he refused. Having anything to do with liquor was strictly against his principles. He did get a license and a permit to sell kerosene as fuel for cooking in portable stoves, which he gave to Kedar, who started a retail business selling kerosene oil. The permit allowed him to buy wholesale directly from the supplier. Because kerosene oil was always in short supply and sometimes not available, Kedar sold pungent mustard seed and rapeseed oils, to supplement his limited income. These items were very much in demand but not rationed.

By the time Suri turned fourteen in May, he had a part-time job filling in as relief from time to time for his brother, a job he would keep for the next year. While he finished high school, Suri also worked for Shahji. The teenager was often the one to close the shop in the evening, add up the sales receipts for the day, and balance the cash. Shahji trusted his young son with all this responsibility and even had Suri fill in for him on breaks. The better Suri became at taking care of business, the more he was relied upon. His skill was noticed by others as well. One day his father's business partner asked Suri to fetch some important papers. The man was startled when the boy delivered the documents to him the next day without being reminded.

He complimented Suri, saying, "I did not expect such reliability from a kid!"

The last year of high school was a turning point for students in India. Suri was faced with some serious decisions about his future career prospects. After tenth grade, the norm in that part of India was to enter two years of intermediate studies, similar to a junior college, which prepared students either for university or the working world. For the Sehgals, it was still important that Suri go on to college.

The educational system in India under the British had been skewed toward liberal arts rather than the sciences and professional training.

But the only job options available after graduation in liberal arts were government administration, the railways, law enforcement, or to attend one of the military academies. Joining the army, especially at officer rank, was considered glamorous. Several of Suri's cousins took that route.

Everyone in the family had been convinced all along that Suri would become a civil engineer. But to do so would have required that he go to college in a different city, which was financially impossible. Considering the family's severely reduced economic circumstances, a degree in engineering was now an unrealistic dream. Suri would stay in Amritsar and attend Hindu College. The family was still essentially in survival mode and education continued to be a high priority.

Both Suri and Padma graduated in the spring of 1949 from their respective high schools, Suri getting first division and Padma second. Suri always believed that if his sister had had the same freedoms and opportunities he had, she would have certainly secured first division. However, Padma was the first girl in the family to finish high school, and she went on to college right away, studying liberal arts at a government college for girls.

By this time Shakuntla was enrolled in Hindi language school for higher studies, a degree that did not require the English language. She majored in Hindi and would eventually earn the equivalent of a bachelor's degree.

Shahji continued to nag Kedar about going back to school, but he took a government job in 1949 and discontinued his business. It would be a couple more years before he could be convinced of the value of higher education, particularly for displaced individuals with few resources.[14]

The transition from high school to Hindu College for intermediate studies in Amritsar was not easy for Suri. All his schooling before then was in Urdu or Hindi. Now all lectures, exams, and science textbooks were in English. The language challenge eliminated many students who couldn't keep up. Suri had to develop his English comprehension

---

14. Kedar eventually completed an honors program in the Punjabi language from Punjab University with first division grades, and later completed college and a master's degree in history, attending night school while working during the day.

to follow what was taught in the classroom. He was more easily able to master the science subjects taught in English, but his spoken English, as was true for most Punjabis, was not comparable to his reading ability. Students rarely spoke English in conversation.

Suri was still spending quite a bit of time helping out with his father's business. Inevitably, his grades started to reflect that time away from schoolwork. After the intermediate final exams, he made the decision on his own to make a complete break from Shahji's business and focus solely on his studies. Not surprisingly, Shahji was fully supportive of his son's decision. Suri's grades improved dramatically after that, and his father was very pleased.

Shahji's business was well established by this time. Again he was in demand for his leadership skills in politics and community work, and now he was sought after for a leadership role in the private sector. Within a year after launching his business, he was elected president of the retailers association. The Sehgal home was again becoming a center of bustling activism as he revived his political and community involvement.[15]

On Christmas Day 1951, while Suri was studying at Hindu College, he was walking in the afternoon with his youngest uncle, Shila's brother Satwant Singh, who was only five years older than Suri. Satwant Singh wanted to purchase a pheasant to take home for dinner. On the outskirts of town, on the road heading toward the cantonment, they passed by a Catholic church. The two walked inside the church to have a look.

Suri had only become acquainted with the Christian holiday of Christmas after living in Delhi when he was thirteen, and this was the first time he had entered a church of any kind. The church service had just ended. He noticed that the sanctuary was impressive, clean and well maintained. The people there were a mixture of British and Indian. A friendly British lady greeted them.

Suri thought it was fantastic to be there and not thrown out, but

---

15. Shahji divested his share of the flour and sugar business to his partner within five years due to the increased demand for his attention to association duties and his political and community work. He remained the president of the retailers association for the rest of his life.

actually feel welcomed. After all, they were strangers, and his Sikh uncle wore a turban.

The kindly lady invited Suri and Satwant Singh to her nearby home for high tea. They strolled there together, and she showed them around her beautiful garden before tea was served inside the house. Several helpers offered big slices of a scrumptious cake. The tea was wonderful. Suri had never been inside the home of British people. This was a novel experience on a festive holiday.

His day of firsts continued after Suri and his uncle thanked their new friend and headed back to Satwant Singh's home, purchasing a pheasant from a street vendor on the way. At home, his wife cooked the pheasant and Suri joined them for dinner, eating meat for the first time in his life. He found the meal delicious.

About five years after Partition, Suri heard that the border between India and Pakistan was going to be open temporarily for two or three days, during which time a big cricket match would be held in Lahore. No travel visa was required to enter Pakistan.

The border was less than fifteen kilometers away, so Suri and a few of his college friends decided to attend the cricket game. They took a bus to the border checkpoint and walked across into Pakistan where they caught another bus to Lahore. The young men ended up standing outside the crowded arena rather than actually gaining entrance to the game because many thousands of others had the same idea that day.

But Suri was far more interested in seeing the university campus in Lahore, anyway. So the friends headed to the extraordinary Punjab University compound and stood in awe of the Gothic-style buildings, the impressive red brick walls, and the stunning grounds. The sense of history was palpable. By now, Suri was as enamored with education as his father was.

The young men ended up having a meal in a restaurant where Suri tasted beef. He was not particularly impressed with the food, but the event was memorable in that it was another first.

During his two years of intermediate studies at Hindu College, Suri was exposed to a basic science curriculum akin to premedical coursework. He took a liking to botany in particular. The Punjab

University Botany Department, which in pre-Partition days was part of Punjab University at Lahore, was now located at Khalsa College in Amritsar, offering undergraduate and graduate degrees, including a PhD in botany. Though the Punjab University buildings were in Pakistan, the Hindu and Sikh faculty had relocated to East Punjab. The department rented classroom space from Khalsa College and converted a couple of verandas into laboratories where graduate students and postdoctoral fellows did their research. The botany honors students were considered students of Punjab University, a tenant of Khalsa College.

After completing intermediate studies, Suri enrolled in the honors track, analogous to majoring, in botany at the Punjab University Botany Department at Khalsa College. What Suri learned at this premier institute of higher learning affected his academic aspirations.

Khalsa College was west of the city center, a little over three kilometers from the Sehgal house. Suri either walked to classes or took a bus. When he got a summer job with the election commission during voter registration, he earned enough money, 50 rupees (about $10), to buy a used bicycle. Once he earned a little more money, he sold that bike and bought a new one for 125 rupees. That was a lot of money, but living at home while going to college was much cheaper than living in a college dormitory, which he could not afford anyway.

Suri enjoyed his three years in the honors program at Khalsa College. The group was small, and they all got to know one another well. Two courses impressed him the most: mycology, the study of fungi, taught by Dr. K. S. Thind; and cytogenetics, the study of cells, taught by T. N. Khoshoo. Both men were relatively young professors who would go on to have prominent careers. Both were highly approachable and easy to engage in conversation, unlike the other professors who sat in their ivory towers, avoiding contact with the students. Dr. Thind, who had just returned from the University of Wisconsin with a PhD in plant pathology, was helpful and supportive, stimulating Suri's interest in going to the US to pursue his PhD—though he did not know how this would ever be possible.

After World War II and post-Partition, the Ivy League universities in the US were considered equals to Cambridge and Oxford, the standards of excellence in the UK. Most of the science magazines, journals, and textbooks in Suri's college library came from the US. All the authors were Americans.

For three weeks each summer of the honors program, Suri and thirty or so students from the Botany Department were taken to the Western Himalayas for hands-on experience studying the flora in the region. The college rented space in the Himalaya Club in Mussoorie and provided a cook to prepare regular meals. The students hiked with the professor, identifying and studying plants as they went, sometimes walking in the evenings as well. If it was rainy, they set up a lab to study indoors during the day.

Suri completed a bachelor of science degree in 1955 at the top of his class, referred to as "first class first." He was awarded a scholarship for higher studies, so he stayed at Khalsa for two more years to complete a master's degree in science. During these two years, his research topic took him back to Mussoorie, also known as the Queen of Hills, for several months at a stretch. Suri was happily immersed in studies that held his full attention, and his parents were very proud when he again secured a first class first.

Hoping to explore scholarships and the potential for admission to a PhD program, and to see what job opportunities might be available in his field, Suri went to Delhi in 1957 to visit the Indian Agricultural Research Institute (IARI). Suri had good grades, but he knew that good grades were not enough. Contacts mattered, too, and he hoped to make some. While visiting IARI, Suri learned more about a Rockefeller Foundation team of three eminent scientists who had visited India in 1951 and how, as a result of their visit, the Rockefeller Foundation and the government of India had signed an agreement in 1956 to strengthen the agricultural research system in the country. The agreement to help revamp agriculture in India from a colonial system to a modern land grant college system was exciting news. Suri knew that the land grant system had had a big impact on American agriculture. The future in this field looked promising.

Suri had heard stories about the scientists involved in these efforts from graduate students and postdoctorate fellows, some of whom had studied in the US. Among the experts was a professor from Harvard University, Paul C. Mangelsdorf, who was highly respected as one of the world's leading authorities on the origin and evolution of maize (corn). Corn was already big in America and was now becoming an important crop in India as well. In the wheat belt of the Punjab, corn had previously been raised only for animals. The green stalks were used for fodder and roughage.

Since Suri had majored in cytogenetics, Mangelsdorf was definitely the person for him to know. After talking with friends at IARI, Suri was sure he wanted to do graduate work with Dr. Mangelsdorf and pursue a PhD in the American university system.

While attempting to contact Mangelsdorf to explore this possibility, as a contingency plan, Suri applied for two IARI scholarships he read about in the paper that paid Rs.150 (about $30) per month. He also applied for a job as a lecturer in botany at Deshbandhu College in Delhi.

In spite of his excellent grades, Suri was rejected for the scholarships at IARI and was terribly disappointed. But this rejection turned out to be a critical turning point in Suri's life that he would one day count among his greatest blessings. Had he been selected at IARI, his future would have been very different. Only in hindsight would he appreciate that.

Suri accepted a teaching offer from Deshbandhu College which paid 250 rupees (about $50) per month plus perks—a pretty good starting salary in those days. This gave Suri a job that was related to his field, as well as an income. He found that he enjoyed many aspects of teaching, particularly mentoring students. He worked hard and his social life was limited, no dating of course. His parents assumed they would arrange his marriage eventually.

Suri's family in Amritsar had expanded. Kedar, Shakuntla, and Padma were now married and having children on their own. Suri's little sisters were now young women. Shahji and Shila remained busy and preoccupied with their work and day-to-day struggles. Life continued to be pretty tough for the majority of people in India. Nehru was still India's prime minister and the Parliament had passed some reforms to address caste discrimination and provide more legal rights for women. Suri's awareness of world events during that time was primarily from newspapers.

Through 1958 Suri engaged in an exchange of letters with Professor Mangelsdorf, who encouraged him to apply for admission to the Graduate School of Arts and Sciences at Harvard University. If Suri was admitted to Harvard, the esteemed professor would consider taking on Suri as one of his graduate students. Suri was ecstatic! He immediately applied for admission to Harvard and a couple of other universities, just to be prudent. While Suri was eagerly waiting for an answer from Harvard, Dr. Mangelsdorf wrote to tell him about the progress of the Rockefeller Foundation's agricultural initiative in India.[16]

Previously, the Rockefeller Foundation had been known for its work in medicine, not agriculture, but it saw the potential in the research of the 1940s and 1950s to transform agricultural production in developing countries like Mexico and India. Dr. Mangelsdorf suggested that Suri contact Dr. Ralph W. Cummings about the Rockefeller Foundation scholarship program in India. Dr. Cummings had been heading the program in India since 1956.

Suri met with Dr. Cummings in his office at Kautilya Marg, in Delhi. Like most Americans Suri had met, Cummings was approachable and easy to talk to. He gave Suri many useful tips on studying in the US and encouraged him to apply for their fellowship, but cautioned Suri that he did not actually qualify for a Rockefeller Foundation scholarship. He did not have the required two years of experience in the agricultural field, nor did he have a job in agriculture that he could return to after graduation in the US.

Cummings offered to write to the headquarters in New York to see if this rule could be waived in Suri's case, but the Rockefeller Foundation was strict about the prerequisites for the scholarship. The decision makers could not be swayed.

At IARI, Suri met another friendly and helpful Rockefeller Foundation employee, Dr. Ernie Sprague, who had been sent to India to work as a corn breeder in the program. During one meeting, Sprague offered Suri his personal copy of the monograph, *Corn and Corn Improvement*, saying casually, "Return it whenever you finish."

Mangelsdorf, Cummings, and now Sprague were brilliant scientists

---

16. The program in India was eventually a huge success, culminating in what later became known as the Green Revolution.

who each took the trouble to guide and help Suri move closer toward his desired career in agriculture. Suri found their easy generosity remarkable. Through such friendly encounters, he developed further admiration and respect for what he saw as an American attitude of warmth and receptivity toward aspiring students. Such kind personal attention from someone in a high position was unheard of in 1950's India.

Suri's elders had told him how the British officers kept their distance from the Indian staff; these aloof commanders, school head-masters, and other elites associated only with their own kind in their private clubs. Theirs was a hierarchical world with the most powerful at the top. These American scientists were not like that at all. What Suri experienced with these men felt like a revelation, "a religious experience without the religion," in his words. "Their important lesson was that knowledge is not restricted to those with political and social rank, but is available to those with the right preparation and motivation."

Suri noticed that even as small a thing as the way these Americans signed their correspondence conveyed a respect that he associated with greater sincerity. Dr. Mangelsdorf always signed off his letters with "Yours faithfully."

Such correspondence in India typically ended with "Obediently yours." The concept of obedience permeated Indian culture. Blind obedience was ingrained in children toward their parents, in women toward men, in the lower castes toward the higher, and in the powerless toward those in power. Gandhiji had confronted that concept with his use of respectful civil disobedience to effect change. Suri was grateful for this new type of respect he was experiencing.

Suri was almost twenty-five in the spring of 1959 when he received a letter from Harvard informing him that he was admitted to the Graduate School of Arts and Sciences beginning in the fall semester. Dr. Mangelsdorf suggested that Suri come earlier, in July, and accept a job as his research assistant. This was an invitation to work side by side with Mangelsdorf, doing hands-on field research, before classes began that fall. Suri could work until October and earn enough money to pay his graduate school tuition—about $1,750 per semester at the time. Suri jumped at the opportunity.

"This was more than great news," he said years later, "It turned out to be the making of my life!"

Suri was still working in Delhi when he received Mangelsdorf's offer, so he didn't see his family's immediate happy reactions. But of course, for his son to be going abroad to advance his education especially delighted Shahji.

Suddenly, countless things had to be done quickly. Suri needed to obtain a passport, US visa, and vaccinations; find money for travel and arrange flights; and tie up loose ends in Delhi and at home. These tasks required him to run around frantically to get everything completed in time. Amazingly, he was ready to take off in the first week of July.

Suri's proud father covered his travel expenses. Suri did not know where the money came from, but he knew it must have been a struggle for the family to pool the money for transportation and related costs. They never did move into the house they were allotted after the Partition. When their friends finally vacated the house, Shahji sold the house and land and invested the money in his business. It would be many years after Suri left home that the Sehgals would finally move into a more spacious home on Hukam Singh Road.

The day Suri left Amritsar on the train, his whole family came to see him off. They expressed their happiness that he was reaching his goal in life and covered him with garlands made of yellow marigolds.

Suri's travel route from India to the US entailed multiple complicated steps that included eight separate flights. All his travel arrangements for hotels, meals, and transportation from the airport to the hotel and back at each location were organized by the travel agent, so Suri didn't really know what to expect. There were all these stops and layovers, new countries and strange cities. The long voyage would take several days from Delhi to Karachi, to Beirut, to Paris, to Brussels, to London, to Reykjavík, to New York, and finally to Boston.

As Suri's train pulled away, he was excited yet apprehensive about what lay ahead. He had never before stayed in a hotel, much less flown in a plane. Whenever his family had traveled on business or vacation, they had always stayed with relatives. Now, he followed the instructions of the ground staff in each airport.

Every overnight stay in a new city turned out to be fascinating adventure. Suri's first night at a hotel, in Karachi, became a challenge with such a simple thing as taking a shower. He was not used to hot running water. He was baffled when the water from both faucets came out hot, apparently due to bad plumbing. Adapting to circumstances, Suri ended up taking an uncomfortably hot shower.

Eating dinner with a knife and fork was another challenge, since people in India ate with their fingers. Before Suri left India, his uncle Kartar Singh showed him how to use the utensils, but Suri still struggled awkwardly.

An Indian couple was traveling with a similar itinerary on their way to London. They did not know a word of English and stayed close to Suri, who helped them fill out landing and embarkation forms at each airport and came to their aid in answering questions at immigration and customs. In Beirut, the restaurant at the hotel where they were staying for one night served spaghetti for dinner. Neither Suri nor his two companions had a clue how to consume this strange new dish. They went without much to eat that night.

But there were continuing moments of wonder along with the complications. The next morning, on a short stroll, the three from India were delighted to see the beautiful Mediterranean Sea with its deep blue waters.

In Paris, the Indian couple was still hanging onto Suri when he went to get something to eat in the Orly airport restaurant. They were baffled to see so many white people. There were crowds of them! They had never seen so many in one place. And some of these people were especially tall. The food served was very different as well, and much of the work was being done by women!

Suri noticed beautiful young salesgirls at the duty-free counters in the transit hall. Only men staffed such counters in India. At home, young girls were kept in the house, and didn't even work in offices. Here in Paris, a young woman was managing a shop in the center of the airport!

Seeing women for the first time doing jobs typically done by men was startling, a huge contrast to India. And the women working in the airport shops and stalls were beautifully dressed in outfits that showed their legs! Suri had never seen women's legs exposed this way. The women had stylish hairdos of varying lengths, not the long braids worn by women in India. He saw no head scarves or dupattas (long scarves) on these women, nor the burkas he was used to seeing on Muslim girls. Suri dared not interact with any of the women in the shops; he was too nervous to look directly at any of their smiling faces. Instead, he looked away bashfully.

In Brussels, Suri was impressed by the vast amount of glass used in the construction of the main hall of the airport. In London, he rode on an escalator and used a coin-operated Coca-Cola vending machine for the first time. All of these things were utterly novel, and Suri marveled at the astonishing sights. In a big hall in the airport, duty-free goods included watches, pens, liquor, perfume, cameras, things scarce back in India. The idea of owning an Omega watch was only a fantasy. Suri's most precious possession at that moment was his simple Tissot watch. He dreamed of someday owning a camera and perhaps a Parker pen!

Suri arrived at Reykjavík Airport late in the evening. When he stepped off the plane, he experienced a type of cold that was completely new. The temperature had been over 38°C (100°F) when he left India. Now in Reykjavik, it was about 10°C (50°F) with a cold wind gusting.

He finally arrived at New York's Idlewild Airport (now JFK) early in the morning. After passing through immigration and customs, he boarded a bus to LaGuardia Airport for the final flight to Boston. The bus was beautiful and spacious, with comfortable seats and air-conditioning! Suri had never before seen such an impressive vehicle.

At LaGuardia, when automatic doors opened miraculously as he approached, Suri said to himself, in happy surprise, *Here the doors open by themselves to welcome strangers . . . this must be a wonderful country!*

CHAPTER 7

Uprooted

In Central Europe, a few thousand miles from the Punjab, Edda's story begins in the Silesian Lowlands that are now part of Poland. Edda Jeglinsky could not recall much about her life before she and her family became refugees in the traumatic evacuation of the German population from Silesia when she was barely three years old. Her earliest memories began ten months later in the fall of 1945 in a small village in Bavaria.

Edda's family origins on both sides had deep roots in the Silesian soil. Her father's family came from Breslau (now Wrocław, Poland), a town that was pulled back and forth between different nations many times as it became a commercial, intellectual, and manufacturing hub in Central Europe. Edda's mother's family was from Festenberg (now Twardogóra, Poland), a small town about forty kilometers out in the country.

Silesia's borders changed often before the unification of Germany in 1871. The region had at different times been German, Polish, Prussian, Bohemian, and Austrian. Grand dukes and Habsburg kings had ruled there. The population had been intermarrying for generations.

By the end of the nineteenth century, three quarters of the almost five-million inhabitants of Silesia were German, and one quarter were Polish, particularly those who lived east of the Oder River. Czech and Jewish communities were scattered throughout Silesia, in cities and all through the hinterlands.

Edda's maternal grandfather, Max Wiorkowski, was the oldest of six children of Polish parents in the town of Festenberg. His mother, a woman reputed to be quite beautiful, had become a widow early in life. Max did his best to help support her and his five younger siblings.

As a young man, he had served in the Kaiser's Special Guard, an elite group known for requiring "flawless" men who were tall, strong, and well composed. The guards wore elaborate crisp white uniforms and pointed helmets and rode white horses.

After Max's military service, he was approached by a master craftsman to learn the furniture trade in Festenberg—a town known for furniture making. Always looking out for his family, Max would only agree to begin the apprenticeship if the master would also train Max's brother, who was partially disabled with hearing and speech impairments. The master agreed and both brothers began training in the traditional German way over many years, first as apprentices, then as journeymen, before becoming professionals. Max went on to earn certification as a *Meister*, which allowed him to hire and train his own apprentices.

Max was a married man by the time he was a master furniture maker. He took out loans to begin his own furniture factory in Festenberg. He was hardworking and tough on his apprentices, expecting the very best of them, as he did from himself. But soon after opening his factory, Max's wife became very ill and the medical bills were hefty. He was forced to take out more loans. Tragically, Max lost his wife to lung disease shortly after the birth of their first child, a daughter named Erna, on February 25, 1910.

Max's late wife's two unmarried sisters agreed to care for the infant, but the women sat and knitted throughout the day to earn money, paying scant attention to baby Erna. When Max visited and found his precious little one neglected and malnourished, he scooped up the pitiable girl and took her home. He burned all the baby's belongings, including the crib because it was so filthy. Max began an earnest search for a new mother for his child.

A family in the neighboring town of Goschütz agreed to the temporary care of baby Erna. Max had a weekend job as a waiter in a popular restaurant night spot in that town. At a dance gathering there, he met an attractive and outgoing woman named Ida. At twenty-eight,

Ida was two years older than Max. She had an enviable position working in the castle of the count of Goschütz as first lady-in-waiting to the countess. Max and Ida had a fairly swift courtship. From the time Max took her to visit his daughter, kindhearted Ida fell in love with baby Erna. She gave up her job with the countess to marry Max, took over the care of his baby, and nurtured the child back to health.

Though Max worked hard for many more years to try to get out from under the heavy financial debt he carried, he and Ida had two more children: a son, Alfred, born February 14, 1913, and a daughter, Margarete, born June 25, 1919.

After the Great War, which had ended in 1918, hyperinflation caused money to be worthless. Max struggled to survive financially; furniture orders were few. He and Ida and the children lived as simply as possible, relying on a big vegetable garden, fruit trees, and some goats, chickens, ducks, and geese. Each family member had one pair of shoes that had to last for a year. Max was pragmatic in the face of his family's financial limitations. He was as tough on his children as he was with his apprentices. When Margarete started school and needed glasses, Max responded, "In our family, nobody needs glasses."

Margarete felt somewhat alone as a child growing up with a sister nine years older, a brother six years older, and parents who were working so hard all the time. She felt closest to her brother Alfred, who teased her a lot. She enjoyed his attention. Her best friend was a classmate named Charlotte Jeglinsky, whose parents, Günther and Emma, had moved their family from Breslau to Festenberg during the Great Depression that was devastating all of Europe and most of the rest of the world by that time. Günther had opened a meat shop, hoping that life in the country would be easier than it had been in the city.

Charlotte's brother Heinz, four years older, adored young Margarete Wiorkowski. His persistent attention was successful in gaining her favor, though she was barely a teenager by the time the Jeglinskys decided to return to Breslau. Their meat shop had not been a success. Cosmopolitan by nature, Günther and Emma felt ill at ease anyway in a small town where everybody knew everyone else's business. They were not comfortable, nor did they feel accepted, during their sojourn from city life.

After he and his family moved back to Breslau, Heinz kept in touch and visited Margarete in Festenberg. His visits became less frequent when, seeking adventure at age sixteen, Heinz joined the merchant marines in 1931. He continued to visit Margarete whenever he could between long sea voyages. Her loneliness increased when her brother Alfred completed his apprenticeship at Max's furniture shop that same year and enlisted with the Reiter Brigade (cavalry) in Breslau. She didn't see her brother very often after that.

When Margarete turned fourteen, school was a luxury her family could no longer afford. Against her wishes, she had to quit school and go to work to help support her family. She had been a good student and loved school, particularly sports, and hoped to become a gym teacher. Instead, her father arranged a three-year apprenticeship for her in a clothing store, which she was obliged to take. The agreement was a promise made between Max and a friend of his from the local fire department where he worked as a volunteer. Margarete's boss, as required in similar apprenticeship agreements, sent her to trade school one or two days each week, and her salary would increase a little bit each year.

When Margarete started working in the clothing store, she decided the time had come to cut her long hair for the first time. She'd always worn it in two thick braids to her waist, just as all young girls in the town did in those days. Margarete found such long hair to be more of a burden, especially in the winter months when it took so long to wash and dry. Girls in the bigger cities, such as her friend Charlotte in Breslau, cut their hair shorter and wore stylish bobs.

Margarete did not consult with her parents about her decision to have her braids cut off. Later that same day, one of her father's friends in town ran into Max and mentioned seeing Margarete's new haircut. Max stormed home to see for himself. He was furious that his daughter would do such a thing. He held the view that long hair was a woman's most precious ornament. He refused to speak with Margarete for six full months. The teenager was not happy that her father was angry, but her new spirit of independence as a young working woman left her with an attitude of "Oh well, too bad." Max got over it eventually.

Overcoming her considerable disappointment at being forced to leave school in the first place, Margarete focused her attention on becoming an excellent employee. She did a bit of everything—sales, bookkeeping, collections, sweeping up—and she did each thing well. Though her boss valued her work ethic and easy way with the customers, outwardly he was demanding, stingy, and temperamental.

During this time Heinz Jeglinsky continued to come and go as a merchant marine. But the relationship basically tied down Margarete as a very young teenager, at least as far as boyfriends were concerned. Heinz was her first, and only, love interest. Though the two were not formally engaged, Margarete, by age sixteen in 1935, was fully committed to Heinz, and there was no turning back.

Adolf Hitler had become the German chancellor in 1933. As a teenager, Margarete was aware of the rise of Nazism in her country, though she never attended any of the meetings where thousands assembled to listen to the chancellor speak. She had the general impression at first that what was happening in Germany in those early years was probably a positive thing, particularly for young people. New clubs for boys and girls were formed, and the emphasis on youth helped to instill some much-lacking pride in the German people for the first time since the Great War and the Depression.

Joining the Hitler Youth was made compulsory in 1936 for all German children ages ten to eighteen. Seventeen-year-old Margarete joined, as required, and went on some outings with the youth groups. They sang folk songs and played sports, but when she noticed politics creeping into the activities, she shied away from involvement.[17] Neither Margarete nor her family had any interest in what they considered the circus around Adolf Hitler—the person, his speeches, and the large assemblies in the town square where everyone had to do the "Heil Hitler" salutes.

---

17. When Hitler came to power, he abolished all the youth groups that had become popular in the 1920s in favor of Hitler Youth, which was designed to prepare boys for military service and girls for motherhood. Margarete was nineteen when it became mandatory for kids ten to eighteen to also attend evening meetings that included vigorous indoctrination into Nazi notions of racial purity. http://www.historylearningsite.co.uk/hitler_youth.htm.

Margarete's father used to say, "This is just rubbish!" and that was that, as far as Max was concerned. The family basically dismissed Hitler as a *Spinner*, a German word for a person who doesn't make sense, is mixed up or crazy in some screwball way, a person with grand ideas that no one would really believe.

The family did not anticipate or predict that their country's leader might actually be dangerous. For Jews and others to be systematically targeted in more and more ways made no sense to the Wiorkowski family. The only Jewish family Margarete knew of owned a clothing store in Festenberg, a competitor of the store where she worked, which was considered to be very exclusive. People shopped there when they wanted something special.

When her three-year apprenticeship was completed, Margarete quit her job in the clothing store. Fed up with her inconsiderate boss, she was immune to his pleadings for her to stay. She had no further interest in working in clothing sales. However, her knowledge of quality goods and her natural flair for dealing with people turned out to be important assets in various ways throughout her life, particularly during the difficult times to come.

Margarete made the decision to leave her family in Festenberg. At seventeen, she moved to Breslau to take an office job working alongside her best friend, Charlotte. Heinz and Charlotte's parents invited Margarete to live with them in their apartment, and her parents approved.

Margarete loved living with her future in-laws, with whom she got along well, and city living in general. Though it was difficult for Margarete to see Heinz so infrequently, there were lots of things to do in and around Breslau in her free time, such as dances, concerts, and theater. She and Charlotte were young girls enjoying relative independence in the big city. Margarete had fun as a young working woman in a business environment. She was fast and efficient at typing and enjoyed office work. Her work friends were a collegial group, participating in regular office parties, group activities, and ski trips. She had respectful bosses and sometimes traveled for work with her boss or with colleagues. Life in Breslau was even better for Margarete when her brother Alfred got married at the end of 1937 and he and his wife Gertrude moved into nearby army housing.

A sudden wave of violence broke out on November 9, 1938, in hundreds of cities and towns throughout Germany and its annexed territories at the same time. Two days of hostilities were focused on Jewish homes, synagogues, and thousands of Jewish businesses. Store windows were smashed, shops were looted and destroyed, and synagogues were burned to the ground. Nazis in civilian clothing, along with stirred-up Hitler Youth members, carried out the well-organized assault. However, the German news provided an official report that these were all spontaneous acts. The attack was referred to as *Kristallnacht*: "the night of broken glass."

Margarete and her family at home were shaken by what happened. She was especially upset to hear about the destruction of the finest clothing shop in Festenberg. But she and her family assumed the government-controlled media stories were true, that the violent behavior was some sort of mass hooliganism.

It would be some time before she understood that Kristallnacht was part of a much larger and more insidious plan Chancellor Hitler had been engaged in from the moment he took power. His intention to rid Germany of certain types of people and create a German master race involved ever-increasing terrorism that resulted in "the systematic, bureaucratic, state-sponsored persecution and murder" of approximately six million Jews and various other groups Hitler considered to be inferior: all people with disabilities, Slavic people, gypsies, homosexuals, communists, socialists, union members, and on and on.[18] The Holocaust had begun.

Germany invaded Poland in September 1939, resulting in the onset of World War II. Heinz was drafted and assigned to U-boats. Until then his life as a merchant marine had entailed long sea voyages with intermittent and infrequent breaks; but now, as a member of the German military, he had regular shore leave with more frequent opportunities

18. "Introduction to the Holocaust." http://www.ushmm.org/wlc/en/article.php?ModuleId=10005143.

to spend time with Margarete. He learned that he could have an even greater amount of leave time as a married man. Setting a date for their actual marriage, after their six-year engagement, was happy news for both families, especially in contrast to their country's involvement in an expanding war.

Heinz and Margarete's wedding was held at her parents' house in Festenberg on December 28, 1940. Margarete's work colleagues arrived by train from Breslau for the celebration. There was an all-night party with music and plenty of good food and drinks. Friends and family talked about that occasion for a long time.

Margarete continued to work in the same office as her best friend, and now sister-in-law, Charlotte, and to live with her husband's parents. The workplace had changed as a lot of the men Margarete had worked with were now in the German army, and throughout Silesia more women had joined the overall workforce. Everyone knew people who lost family members in the war.

Heinz was on duty, mostly in Scandinavia. Margarete sometimes traveled by train to see him when his shore leave was in North Sea port cities, such as Lübeck or Bremen. The only conversations that he ever engaged in with his wife, related to the war, were about the things he experienced on shore leave in Norway and Denmark. He frequently commented on how impressed he was by the honesty of the people in Denmark. He said a person could leave a bicycle unlocked along a road there, and it would never be stolen.

Being part of the military, Heinz had it pretty good as far as Margarete could tell. He didn't experience the shortages of food that people did in Silesia. He brought back cheeses, cans of sardines, and other items that were no longer available at home.

When Heinz found out that he could have additional time off with his wife if they had children, the couple decided to have a baby.

The birth of Heinz and Margarete's first child, Edda Gudrun Jeglinsky, was attended by nuns in a Catholic hospital in Breslau on Christmas Day in 1941. Margarete had hoped that her baby would arrive before midnight, like the Christ child.

She later told Edda, "But you were fifteen minutes late!"

In fact, Margarete had been engaged in a more serious tug-of-war on Christmas Eve regarding the timing of her baby's birth. The nuns in the hospital, similar to nuns in schools, lived within very regimented schedules. Christmas being their most exciting celebration of the year, they were quite eager to celebrate midnight Mass at the church, not in a hospital.

Midnight was fast approaching, and Edda's impending birth was interfering with the nuns' plans. They did their best to convince Margarete to delay the birth, insisting that there was plenty of time.

But the baby was already overdue by several days, and Margarete had a gut feeling about her child's birth. In her concern, she was rather insistent about not delaying anything. As it turned out, her instincts were right on target. The umbilical cord was wrapped around the baby's neck. If they had waited to appease the nuns, the baby might have suffocated. The nuns missed midnight Mass in the church, and Edda was delivered safely right after midnight, Christmas morning.

Mother and child stayed in the Jeglinsky apartment in Breslau for a while, but Max wanted Margarete and his new grandchild to move home to Festenberg. He pointed out to Margarete that life in the country would be better for the child than living on the high floor of a city apartment building in Breslau with the Jeglinskys. Besides, life was safer in the country. The cities were being bombed.

Erna had moved home, too, after her husband Robert was drafted into active service. Alfred and his wife Gertrude and their baby, born the previous April, stayed in Breslau where Alfred was still stationed in the cavalry.

Margarete agreed with her father. She and Edda moved back to Festenberg.

Other than the regular sight of armies marching through to the Eastern Front, the effects of war in general did not cause much disruption to everyday life for the family in Festenberg. Their basic needs were met. They had their own vegetables, fruit, and meat, so food wasn't a problem. Max worked hard and maintained his connections in the local business community, socialized on the weekends, enjoying a drink with friends, and attending the town dances. He was a sought-after dance partner for the ladies whose men were away at war. He danced with everybody, and Ida didn't mind. But his soft side was most evident around little Edda. He showered attention on the baby.

Edda's family felt relatively secure at home throughout most of the war. Max and Ida focused their loving attention on their grandchild. Margarete told Edda again and again, over the years, how close Edda and her *Oma* and *Opa* were during those first three years of her life. Margarete said Ida was an angel to Edda, and Max doted on the little girl. Margarete joked with her, "You will never have it so good again!"

Max's furniture business was finally picking up. He was almost out from under the huge debt. Furniture orders began coming from clients in Breslau who paid well.

But the horrors of World War II were not too far away from their somewhat-safe haven in the country. Sad news came frequently. Heinz's father, Günther, died soon after Margarete and Edda moved to Festenberg; and word came that Heinz's half-brother, Walter, was killed while on active duty.

Over those three years, Margarete and Heinz scheduled their time together during his shore leaves in places such as Zingst, an island in the *Ostsee* (Baltic Sea). Heinz wanted to see his wife and daughter whenever he could. The summer Edda was eighteen months old, the three were vacationing there on the beach—far from the realities of war. The summer season in Silesia was typically so short that any time in the sun was treasured. All along the shoreline, vacationers were dozing on lounge chairs with cloth canopies. Margarete and Heinz lay back in their beach chairs while Edda played happily in the sand. When Margarete finally noticed exactly what Edda was doing to entertain herself so busily, she was surprised to see that the toddler had collected the sandals and shoes from all the other people along the shore and created a big pile of footwear right in front of Margarete. Edda said proudly, "Mommy! Shoes!"

A charming and outgoing toddler, Edda was comfortable going up to almost anyone. Her easygoing ability to entertain herself for hours at a time came in handy in an emergency one evening during a visit with Heinz in Bremen, when he was suddenly called back to his ship. The landlady of the house where they were renting a room had allowed Margarete to do her husband's laundry, and he now needed his clothes before the ship pulled out.

Margarete couldn't carry both the laundry and the toddler on public transportation back to Heinz's ship before he left. Under the rushed and difficult circumstances, she had no choice but to leave Edda alone in her room for what could be more than an hour, though the landlady was elsewhere in the house. Margarete quickly instructed Edda, "Stay in this chair, and don't go anywhere until I get back," before she ran out the door with Heinz's clean laundry.

When Margarete returned a couple of hours later, she rushed into the house to find Edda sitting quietly in the same chair, happy to see her mother. She didn't appear to have moved even an inch since her mother left, or to be at all bothered by the time she spent alone. Margarete was at once relieved and amazed.

The Allied Forces were closing in on Germany in early 1945. Hitler's reign was collapsing, and the city of Breslau was key to the southern invasion route into Germany. The largest city in the former eastern territories of Germany, Hitler had designated Breslau as a *Festung* (fortress), one of seven cities to be held at all costs, "to be defended to the last drop of German blood."[19] Alfred was drafted to defend Breslau right after that announcement. Margarete feared for her brother's safety.

The *Gauleiter* (the regional Nazi leader) for Lower Silesia, an SS man named Karl Hanke, did not permit or even reveal the possibility of evacuation from Silesia until the Russian army was closing in on Breslau by way of Festenberg in mid-January. He gave the order that women, children, and old men must leave Lower Silesia. They had twenty-four hours to leave for German-occupied Czechoslovakia.

The weather was particularly severe that winter; the temperature was often -20°C (-4°F). Max sprang into action and managed to get a wagon with horses. Margarete grabbed her fur coat and what few

---

19. *After the Reich, The Brutal History of the Allied Occupation* by Giles MacDonogh, Basic Books, 2009.

necessities she could carry: some photographs, bars of soap, and a little potty for Edda. Abandoning their home, Max loaded up Ida, Erna, Margarete, three-year-old Edda, and a few neighbors, and they set off together in the cold, joining the rest of the fleeing German population from Silesia.[20]

The family felt particularly upset leaving their animals—goats, chickens, rabbits, a gaggle of geese, and a flock of ducks. Who would feed them? They tried to imagine that someone staying behind might care for the animals. In retrospect, though, they realized their animals were probably butchered by the hungry Red Army that was immediately bearing down on the town.

When the family reached the railway station, people staying behind took the horses. Though the trains were jammed, Edda's family was able to stay together in the same car of the first of many trains they boarded in the coming days, weeks, and months. The train cars had no windows and no heat. Margarete was very glad she brought the potty for Edda. People were so tightly crammed onto the train that some of them could not get to the train toilets. They wet their pants, which would then freeze. A few individuals died of burst bladders in the bitter cold. Many were too weak to withstand the extremes in temperature and the continuing discomfort. Thousands leaving Silesia that winter died from hypothermia in transit.

The trek by train wound through Czechoslovakia, with many stops along the way. The refugees were required to walk long distances back and forth from the train stations and wherever they were told to stay at night. Hundreds marched together for miles in the cold to stay overnight or longer in schools. They slept on straw that was strewn wall to wall on gymnasium floors. Ida suffered with pain in her joints. She had difficulty walking and struggled to climb on and off the trains.

Refugees focused only on basic survival, food in particular. A distribution system was set up for people to receive a small allotment of various food items. Ration cards were issued and used to obtain meager supplies, some bread or potatoes, at stops along the way in the seemingly endless journey. People waited for hours in long lines for their allowances. Everyone was cold, miserable, and hungry.

---

20. Refugees fleeing from further north headed to East Berlin and northern Germany.

Mothers with young children were allotted a quarter liter, about a cup, of milk per child. Every day, in whatever town they were in, Margarete went to the assigned location to receive her milk for Edda. She would return with the precious milk, wait her turn at what was usually a single stove in a large hall full of people, and cook *Grießbrei* (cream of wheat) for Edda. But Edda would not eat the lumpy mush. Margarete tried everything to make the toddler eat the cereal, which she knew was an important supplement to their already limited diet. Margarete cajoled, bribed, even spanked, but nothing worked. In the end, the food was not wasted of course. Someone in the family ate it.[21]

To get through the unending travel ordeal required effort from everyone in the family. Edda's grandparents were a godsend, helping to entertain the child. Erna had always been fairly fragile and lacked the robust health her father enjoyed. She needed help coping with various requirements and discomforts, but she gladly helped take care of Edda.

Margarete, on the other hand, was energetic, dynamic, and inventive in the face of adversity. She always found a solution to any problem, and nothing seemed to get her down. Even though she had no idea how or when she would see Heinz again, she maintained a steady positive spirit as far as anyone around her could see.

As the man in the family, Max was expected to take charge in most circumstances, but he recognized his daughter's talent in dealing with people. She was wise and pragmatic. Max noticed how often others came to her for advice. So whenever they had to meet new people or ask for help, Max told Margarete, "Girl, you go; you can do it better."

Margarete was able to make easy contact with the officials or any other people they came across. She was sure to have far more success approaching strangers with requests than Max might. He was smart enough to know that the frayed officials and public servants were less likely to put up with a gruff old man than a lovely and vivacious young woman.

As weeks and weeks went by, Margarete was determined to find better accommodations for the family each time there was another

---

21. Though Edda retained no memory of her mother's efforts to make her eat the dreaded *Grießbrei*, she continued all her life to avoid any food that was similar in color and texture— any kind of porridge, pudding, polenta, or white gravy. She said, "I don't want to look at it."

lengthy layover. She hated staying in crowded shelters, sleeping on straw in huge halls, with children crying and the endless arguing among desperate people. The spaces were packed, dirty, noisy, and uncomfortable. For Margarete, cleanliness was next to godliness. She could not tolerate the misery her family was forced to endure. Thanks to her ability to seek out and ingratiate herself with local people wherever they went, the family often ended up staying in private homes where they were treated more considerately. People tended to trust Margarete. They could see that she was a decent woman, reliable and clean.

At one point, still in Czechoslovakia, Margarete found good accommodations for the entire summer in the home of a woman whose husband had not yet returned from the war. They were able to live in small but quite adequate private quarters, and for the first time they had dearly coveted laundry privileges. Margarete even learned cooking from her landlady, who was quite talented in the kitchen.

Throughout their journey, Edda was suffering from a terrible rash all over her body, blisters that itched and made her miserable. It wasn't chicken pox, but it was pox-like. The sores oozed, and wherever it oozed, new blisters erupted. If bandages were applied to the sores, they merely broke open when the bandages were removed. There was some speculation that the rash was caused by the straw they slept on at night. The only thing that saved Edda from complete misery was the soap that Margarete had brought along when they left Festenberg. It was a particularly strong antiseptic type of soap for washing clothes. Margarete used it to bathe Edda whenever possible, which brought some relief from the itching.

Finally, in the early fall of 1945, the family reached Bavaria. In this rich agricultural region with adequate food supplies, every farm family was obligated to take in one refugee family. Edda's family was assigned to live in Baldersheim, a rural district of Ochsenfurt, in the northwest part of Bavaria called Franconia. This small village consisted of a cluster of thirty or forty farmhouses, surrounded by their fields. With all the men away at war, many never to return, extra hands were needed to work alongside the farmwomen in the region.

The largest town near Baldersheim was Würzburg, about thirty kilometers to the north. Edda's family left the train in Würzburg, prepared to go on foot to their newly assigned home. They were walking along the side of the road when a truck with American soldiers came by and offered to give the family a ride partway to their destination. The sight of friendly American soldiers, and their gifts of chewing gum for the children, became common in postwar Bavaria.

The US Army had taken on the task of occupying and assisting in the governing of war-torn Germany, a role they would continue with, to some extent, for many years. Most of Silesia was now attached to Poland, and German refugees who left Silesia and tried to return postwar were denied reentry. Displaced refugees were scattered throughout Germany.

How refugees found their loved ones under these circumstances was a remarkable process. In the local town halls where refugees registered, lists were posted of soldiers missing or confirmed dead. There were bulletin boards there and also in all the schools and camps where refugees stopped along the way and people left messages. With those notes, people eventually made contact with each other. During all of the chaos at that time, it seemed a miracle that Edda's surviving family members eventually ended up in the same little town.

Heinz came almost immediately to Baldersheim. He was never held in a labor camp or a prisoner of war camp as most captured German soldiers had been. He found his mother, Emma Jeglinsky, living in a farmhouse with his sister Charlotte, Charlotte's husband Walter, and their baby Rosemarie. Walter was at least twenty years older than his wife and, due to his age and poor health, he'd never been drafted.

Heinz learned through the postings that Margarete and Edda were in Würzburg. He rushed there to meet them. To his surprise, instead of finding his wife and child, he found Alfred's wife Gertrude and their four-year-old daughter, Anita, in an abbey with a community of nuns. Gertrude and Anita had fled Breslau by way of Hirschberg, Silesia, and had spent quite a while in a refugee camp in Regensburg in the Bavarian Forest before reaching the abbey in Würzburg. By now they had learned the tragic news that Alfred had died in March during the Siege of Breslau.[22]

---

22. The Siege of Breslau ended on May 6, 1945, only one day before Germany surrendered.

Heinz brought Gertrude and Anita back to Baldersheim to stay temporarily with his mother and sister's family, which turned out to be where Max, Ida, Erna, Margarete, and Edda were now headed. Within days, Heinz was happily reunited with his wife and daughter.

Edda and her parents were fortunate to end up with a very pleasant family in Baldersheim. Max, Ida, and Erna were given similar accommodations nearby. Where Edda lived, the man of the family was still gone. There were two children. Alfred was around Edda's age, and Beate was two years older. The two girls would become best friends. Not all refugee families were as fortunate. Some refugees were assigned to people who were stingy and resentful of the forced arrangement.

Heinz, Margarete, and Edda lived in the upstairs part of the farmhouse; the host family lived downstairs. The upstairs consisted of a combined kitchen/living room and a room where they slept. There were no mattresses, so they were still sleeping on straw, the only bedding available. The farmhouse had no indoor plumbing, so water had to be carried upstairs to cook, bathe, and wash dishes or clothes. There was an outhouse across the yard. Edda hated going there because, like on many German farms, the doghouse was next to the outhouse. The dog was kept there on a chain. Going to the outhouse was scary for Edda. The dog barked and strained on his chain each time she walked by.

Host families were required to comply with certain regulations. Though supplies were limited, they were expected to provide the refugees with rations of staples such as potatoes, flour, milk, meat, and eggs. Each family received a quarter-pound of butter a month. Edda's family's farm host was a generous woman who sometimes gave them extra flour or milk.

That first fall in Baldersheim, Edda was admitted to a hospital in the nearby village of Aub to be treated for the horrible rash that still plagued her. She spent an entire month in the hospital. Each day, she was given spinach and potatoes for every meal. Slowly the rash began to heal. However, the mystery around what caused such a horrible skin condition was never fully solved. Whether it was a mineral or vitamin deficiency, an allergic reaction to the straw Edda had to sleep on for so long, or something else, Margarete never knew.[23]

All in all, for Edda and her family, life was not so bad. Although everyone was usually hungry, they had survived the war. They were alive and safe.

---

23. The thinking at the time was that Edda might be allergic to red currents—berries often found in Silesia and throughout Germany in backyards in three colors: red, white, and black. Many years later, the same rash returned in a much milder form.

CHAPTER 8

Apples and Potatoes

Extra food and other advantages were possible if refugees helped work in the fields of their host farms. Assisting with the harvest benefited everyone involved. After the harvest, families foraged for leftovers. They collected the tops of wheat, called *Ähre*, not the individual kernels, but the tops that were left behind, *die Ährenlese*. That ritual continued every fall for many years by people all over Germany. The collected wheat could be taken to the local mill and ground into flour. During the growing season, children walked the potato fields row by row and picked the potato bugs off the plants and put them in jars. There was a small reward for each full jar. Edda remembers walking in the fields, collecting overlooked potatoes on the ground.

Along the roads in rural areas were fruit trees, mostly apple. The refugees collected the fallen fruit to make applesauce. There was nothing more special to eat than sweet applesauce on bread. In fact many of Edda's favorite snacks involved bread with something tasty spread on top. From sugar beets, Margarete cooked syrup in a big pot and spread the thick sweet liquid on bread. When the cream was still on top of the milk, Margarete put that on bread with a little sugar for Edda. Sometimes bread was not even needed; when she had the opportunity, Edda would steal a tiny piece of butter and quickly put it in her mouth. Not one thing to eat was wasted, and every bite was appreciated. That period instilled in Edda a lifelong enjoyment of potatoes and good rye bread.

The people in Edda's life post-World War II were primarily focused on survival—hardworking people struggling for their next meal. Money was worthless, just as it had been after the previous war. An informal barter system emerged to trade for the goods they wanted. Margarete traded her fur coat for a radio.

Heinz was handy, an excellent mechanic who could fix anything. He parlayed his services into extra helpings of food for his family. On the U-boats, he had worked skillfully with fine technical instruments and machinery. Now he repaired whatever needed fixing for the farm wives in the area. He could design and create useful items, which he bartered for food. He made wooden sewing baskets cleverly designed to open up accordion-style to hold thread, needles, buttons, everything needed for sewing and mending.[24]

Max raised rabbits in cages, about ten at a time, for extra meat. They were fed potato peels and kitchen scraps, and Edda collected clover and dandelion leaves for them. Each Sunday, Max killed a rabbit for the family dinner. Doting on Edda, as he was fond of doing, Max gave her the little puffy tail to play with and always gave her the cheeks—the choicest, tender part of the cooked meat.

Edda was generally wary around most other animals on the farm. The dog scared her of course; he bit her once when he was loose. A male turkey with a beautiful big tail attacked her when she was wearing a red coat. Like a bull, the big bird ran after her, and she was terrified. Geese were also mean when they had little ones, and their bites were quite painful. There were pigs and big horses that kicked, so Edda remained cautious around most animals.

In many ways, though, Edda was growing up within a cocoon of innocence, shielded by her family from the tough realities of the time.

Her parents did their best to observe as many traditional German holiday celebrations as they were able. A celebration of Christmas started on December 6 with *Nikolaus* (St. Nicholas) leaving sweet treats in a shoe or on a plate left outside the door. *Christkind* (Christ child) brought presents on Christmas Eve and the family would have trimmed a tree with decorations and lighted candles. Christmas Day would have meant church services and a feast—a roast goose dinner, plus cakes

---

24. Margarete had a pretty yellow one, which Edda still uses. See photo.

and chocolate and marzipan. But Edda's first Christmas in Bavaria was a pretty toned-down version of those things. The same was true for Easter.[25] While taking a morning walk with her family on Easter Sunday, Edda discovered colorful eggs that had been placed by Max in the grass and behind trees and bushes in the nearby woods and meadows. This seemed magical to four-year-old Edda who truly believed an Easter Bunny left them there for her.

Margarete had noticed a change in Edda's personality from the time they left Festenberg. Where she had once been an outgoing child, she was now more introverted and shy. She was timid with strangers and hesitant to do new things. Where she once was more active and willing to try anything, now she was cautious, more likely to stand back and watch others. She did learn to swim, which was a little scary, but she otherwise held back from anything too daring.

The host farmwoman kindly invited Edda to eat meals with the family downstairs on many occasions. She was also invited along to their Catholic church and to walk in religious processions, which took place frequently. Edda, Beate, and the other children collected baskets full of flower petals, then walked in front of the procession, sprinkling petals on the road for the priest and nuns to walk on.

Almost everyone in Bavaria was Catholic. Edda's relatives were the only Protestants in Baldersheim. There was one Protestant church, three miles away in Aub, but her family did not go there. This was not a problem, though; Edda did not know she was Protestant. The Catholic customs in Baldersheim seemed perfect to her.

When Edda attended kindergarten, at ages four and five, the teachers were nuns. In that small town, whenever meeting a nun, children were expected to kiss the cross hanging around the waist of the nun's habit and say, "Praise be to Jesus Christ!" Little girls were to curtsy in greeting to any adults, but especially nuns. Little boys had to bow. Edda had no reservations about performing these rituals. To her, it all seemed wonderful. Her friend Beate did those things, so Edda did them, too.

At age six, in the spring, Edda was enrolled in first grade. To prepare for school, Margarete took Edda on a train to Würzburg with

---

25. The traditional Easter celebration was a four-day holiday, with church services and special food, cakes, and sweets.

bundles of old newspapers. Paper was still in short supply after the war, so newspapers could be traded for items such as school supplies. Edda thought this journey by train was an adventure. In exchange for her newspapers, she was given two little silky aprons, one red and one green, the type all Bavarian schoolgirls wore. She also needed a schoolbag (*Ranzen*), not an ordinary backpack but a special satchel for schoolchildren.

The first day of primary school began with a ceremony to mark the occasion. Edda had new clothes, her Ranzen, and she wore one of her pretty new aprons. Each first grader received a cardboard container shaped like a large ice-cream cone, which was decorated and stuffed with goodies: pencils, chalk, and candy. On this festive and proud day, little kids posed with their schoolbags for photos. Edda smiled for pictures with her cousins Anita and Rosemarie.[26]

Since first graders had no paper to write on, each student used a slate and gray chalk, with a sponge and a dry cloth to erase the slate. Every morning, Edda had to make sure her sponge was wet and her cloth was clean. Her sponge was kept in a metal box to stay moist, and the cloth was attached to the slate by a string. Her chalk was thin, and she could sharpen it like a pencil.

Though life was slowly achieving a level of normalcy for Edda's family, there was still no money, little food, very few possessions, and lots of hard work. Bavaria was an agricultural region that would stay at the bottom of the German economy for many years to come. As men returned home to their families in Baldersheim, and other farmers came home, too, extra workers were no longer needed to fill all the jobs. Before long, there was no work for Heinz in or near Baldersheim or anywhere else in Bavaria.

He heard there were jobs to be had in Baden-Württemberg, a highly industrialized state to the southwest. Heinz had an uncle in the state capital, Stuttgart, more than 100 kilometers away, so he took his bicycle on a train to see his uncle and pedal in and around Stuttgart to explore job possibilities. He found the city in ruins. Mountains of rubble were everywhere. In the streets, houses were gone; only some

---

26. In Germany, the start of school was a much bigger celebration than graduation.

walls were still standing. He found no job prospects in the bombed-out city. So Heinz's search extended quite a distance into the surrounding region.

He was finally successful, finding a factory job making precision instruments in Göppingen, about forty kilometers east of Stuttgart. But there was no housing available in Göppingen. The only place he could find was a small apartment in Gingen four towns away, southeast along the Fils River. He would need to pedal his bike the 16.5 kilometers to work every day. But a job was a job, and he was grateful to find work.

Edda and her parents moved to Gingen early in the summer of 1948. The rest of her extended family remained in Bavaria, except for Erna and her husband Robert. By the time Robert finally came home from the war, after being detained, tortured, and starved for more than a year in a Russian prisoner of war camp, he was sick, malnourished, and missing an eye. He and Erna now moved to the county seat, Ochsenfurt, which was still a picturesque, historic town surrounded by fields of sugar beets. Robert was able to make a good living working in the sugar factory, and tailoring on the side.

Leaving friends and family was hard for Edda and her family. Their apartment in Gingen was tiny, but the family who owned the house had three boys, so Edda had friends right away in the new location. Her outgoing personality reemerged a bit playing ball with the boys. A grade school friend named Inge lived close by. She and Edda remained friends all through school.

Residents of the town of Gingen spoke a different dialect than she was used to hearing in Bavaria, and Edda did not want to be different. She wanted to do away with her own Bavarian accent in order to blend in at school. She announced to her mother that she had to learn *Schwäbisch*, Swabian, a different dialect of High German than her parents spoke. She listened carefully to her playmates that summer and paid close attention to word pronunciation. By the time she started school, she had a fairly good handle on *Schwäbisch*. Eventually she spoke like a native and ended up being the only one in her family who ever did.

Edda fit in well at the new school, was a straight-A student, and felt happy in Gingen. She and her friends played ball, hopscotch, and jumped rope. But a lovely gift she received that winter from her uncle Robert threatened her intent to be just like everyone else.

A gifted tailor, Robert made his niece a beautiful navy-blue winter coat with a Navy-style flap in the back and a double row of buttons in the front. The coat was lovely, but the second-grader hated wearing it, simply because no one else had one like it. Margarete made Edda wear the coat that winter anyway, and she was mortified to be seen in it.

That Christmas, Edda was given a doll—something she'd never had before. Max made her a beautiful white doll bed. Margarete made a pillow and blanket for the bed and sewed and knitted clothes for the doll.

Before long, Margarete was doing a lot more knitting. She was pregnant.

Margarete was having a lot of stomach problems, possibly the result of stress. She had remained outwardly calm through all the difficult times the family went through as refugees. But she suffered now when she ate rich foods or too much fat. She was bedridden frequently for periods of time. The doctor recommended hot compresses and a bland diet, sometimes only oatmeal cooked in water. He called it a nervous stomach.

Edda's sister Petra was born a month premature, on Edda's eighth birthday, Christmas Day, 1949. During the pregnancy, Margarete suffered from an infected sore on her lower back, like a deep pressure wound, possibly from being bedridden. The infection was treated with a black zinc cream. By the time Margarete finally saw a doctor, she had a deep abscess. The infected tissue had to be surgically removed and the wound took months to heal. The concern at the time was that Margarete's infection may have affected the baby or even caused the premature birth. The infant had health issues from the start. She lacked the needed enzymes for proper digestion and developed more slowly than expected.

Edda had told her parents that she wanted a sister, but she did not realize that getting a sister would mean a tiny baby. Her fantasy wish had been for a playmate. But Edda loved her sweet little sister and felt very protective of her. She acted as a responsible second mother to the child, especially since Margarete was still recovering from her infection for some time after the birth. Heinz's good-natured mother, Emma, came to stay for several weeks to help out. She slept on a cot in the living room.

Margarete couldn't stay down long. She was a spontaneous person who preferred to remain active. She enjoyed taking care of Petra and the rest of the family, but she found many repetitive things about housework dreadfully boring. So she had Edda do those things that were the most boring, such as grinding coffee or stirring cake batter. At the end of each week, Edda and Margarete baked a cake together—a pound cake or a yeast sheet cake with apples or plums. Edda was the one to stir the simple pound cake for the required half hour.

The summer after Petra was born, her aunt Erna "borrowed" Edda for an extended visit. This was the beginning of a much-appreciated annual tradition. Margarete allowed Edda to go to Ochsenfurt in Bavaria to stay with Erna and Robert for several weeks during the summer break from school. The couple didn't have any children, and they loved Edda. She'd also have the opportunity to play with her cousin Rosemarie, whose family now lived near Erna and Robert.

Edda had the grandest time with her aunt and uncle. They were each small in stature and somewhat frail due to health problems. Edda thought they made an adorable couple. Both treated Edda like a grown-up. While with them, Edda could drink coffee, which she was very fond of. They baked cakes, went to the library, read books, and sang together every day. Robert could pick up and play any instrument—harmonica, mandolin, violin, anything. Edda especially loved it when he played the mandolin. "He was wonderful!"

Robert and Erna both sang in a choir. Erna had a beautiful voice. Some in the family said that perhaps if Erna hadn't been so pitifully neglected and malnourished as a baby, which left her with a slight hump, she might well have been a professional performer, maybe an opera singer. Erna never had much physical stamina. She had tried working in a bakery in Festenberg for a short time, but she just couldn't do the lifting or the standing behind a counter. However, she was good at very fine crocheting, which she did to earn extra money. She taught Edda how to do all kinds of embroidery. Edda prized the gorgeous collection of handkerchiefs with stitched borders that Erna once made for her.

Given all the household responsibilities Edda had at home, she especially appreciated the free time and being able to read for hours on end during those summer visits to Bavaria.

Edda's family moved later that same summer of 1950 to a larger apartment with an extra room, so Max and Ida moved in with them. Ida's health was deteriorating; she was having vision trouble. She had glaucoma, and cataracts in both eyes. She had a sensitive stomach as well, like Margarete's. Ida had to avoid rich foods. Edda noticed how Ida would ask for a tiny taste of something she knew she should not eat, such as potato salad with mayonnaise that was only made on special occasions, then she felt terrible shortly after eating it.

Max, however, was still healthy and strong. He never needed to see a doctor or a dentist. He found work in a furniture factory in town and enjoyed being productive and assisting with the family income.

Unfortunately, Heinz was experiencing some physical difficulties. His job in the factory in Göppingen required him to stand all day, and he developed back problems. He now rode a light motorbike, like a moped, to and from work. Eventually he got an actual motorbike, but back pain forced him to leave the job by 1951. After that, being a salesman was the only job he could find. Heinz sold everything from insurance to vacuum cleaners to gadgets used in restaurants, bars, and hotels. But his back gave him trouble carrying his heavy bag of merchandise samples, so he traded in his motorbike for a VW Beetle.

In 1952 Edda's family, including Max and Ida, moved to Göppingen. The government was constructing four-story, bare-bones apartment buildings with three families on each floor to accommodate the huge population of refugees and displaced persons. The buildings were made to minimal standards with no amenities. There was a bathroom but no bathtub. The apartment had two bedrooms, one for Margarete and Heinz and one for Max and Ida. The kitchen was the room for everything else: bathing, cots for the kids, and homework at the kitchen table.

The common laundry room was shared by twelve families. A reservation was required to use the room to do all the laundry for an entire month at one time, using a washboard.

Clothes were typically changed only once a week. Laundry was separated by color. Whites were boiled in a bucket and stirred like soup, which made the whites very white. Everything had to hang outside to dry, even in winter. On the coldest days, the clothes were hung in the attic. Edda's chore was to hang the laundry on the lines and bring it in when dry. Hanging ice-cold laundry made her fingers ache.

Life wasn't all work and school, though. For fun, Edda played badminton or ball in the street and jumped rope with her friends. Though the family tended to go to bed early, they took walks together in the evenings. On Sundays, it had always been a custom to take long afternoon walks in the countryside or the woods. Their walks sometimes wound up the hillside past the new US Army base.

Everyone was still undernourished and usually hungry. Meals were small. At school, for the midmorning break, children were given free milk, sometimes chocolate milk, and a roll. Edda came home from school at noon to eat dinner, then went back to school for afternoon classes. Evening supper was typically a sandwich and herbal tea.

Margarete was a good cook. The dinners she made were usually basic one-pot meals—potatoes, vegetables, barley soup, and a lot of stews. Meatless bones from the butcher were used to make broth, because meat was very expensive. An infrequent half pound of ground beef had to feed the entire family. Once a week, there was fish or liver. The family had a bigger meal, such as pork roast or chicken, only on Sundays.

Rarely was there butter, only margarine. Fruit was locally grown: apples, pears, and berries. Bananas were a rare treat. Hearing about watermelons from Italy sounded so exotic. The only time Edda saw an orange was as a Christmas gift. Heinz would peel a single large navel orange, and the whole family shared it.

CARE packages began to arrive regularly from the US. Edda unpacked the boxes, which contained several items, including powdered milk (she hated the smell of it), pieces of yellow cheddar cheese the size of a brick, and a canned meat called Spam. She didn't know what that was.

When the family finally had some money, and cash could actually be used to buy things, they sometimes stopped on their walks at a café. Heinz and Max would each have a beer; Margarete and the girls usually had lemonade. The café was where Edda first tasted a German cola called Afri-Cola. Then she tasted Coca-Cola. "Oh my God, it was so exciting to have a bottle of Coke! And the first taste, I loved it!" she remembers.

Oma Emma visited again more regularly, sometimes taking Edda downtown for a special afternoon of shopping followed by a visit to a good *Konditorei* (pastry shop) for cake and coffee.

Though the family was now settled in Göppingen, and finally doing better financially, the difficult years during and after the war had taken their toll on seventy-three-year-old Ida. She needed a gall bladder operation in 1954, which went very well. That day, Margarete sent twelve-year-old Edda to the hospital with a fresh towel for Ida because no soap or towels were provided by hospitals in those days.

When Edda arrived, the nurse said, "I'm sorry, child, you are too late."

During her recovery, Ida's heart had suddenly given out. The nurse led Edda into Ida's room to see her Oma.

When Edda saw Ida lying peacefully on the hospital bed, she immediately recalled the way her mother always described Ida as an angel. That is exactly what she looked like to Edda—beautiful and peaceful, just like an angel. Amid the sadness of her loss, seeing her grandmother looking so lovely made the young girl think of death as a peaceful, natural thing.

The nurse gave Edda back her supplies, and she went home. When she told Margarete that Oma had died, Margarete was devastated. She had fully expected to see her mother well again after her successful surgery.

Max had by then found another job as a furniture maker during the week. He continued to work hard, but now he initiated a new routine: he took Edda, still her grandfather's little darling, to a movie every Sunday. At that time, a movie cost one mark (about twenty-five cents). They enjoyed a lot of German musicals (*Operetten*) together.

Edda's brother Herbert was born at home in that first apartment in Göppingen three weeks before she turned thirteen in 1954. His birth was a big surprise to Edda; she didn't even know her mother was pregnant. Edda never noticed her mother's belly enlarging, probably because

Margarete always wore an apron, as was the custom to protect clothes from needing more frequent laundering. Edda didn't even know what pregnancy was or where babies came from.

Herbert's birth was assisted by a midwife, an older woman who lived in town. She was a hefty woman with an unpleasant personality. She once told Margarete that she did not like children. She owned her own home where she lived upstairs from her family. She took a liking to Max (presumably) and one day offered him a room to rent in her home. He politely declined.

Margarete and Heinz had not planned to have another child, but they were happy with their surprise son. Heinz was glad to have a boy. Max, who enjoyed all children, was quite pleased with Herbert, too. Edda continued in her role as secondary parent now to both siblings. She enjoyed taking her little brother around in a baby carriage to get fresh air. All German children were thought to eat and sleep well if they had sufficient fresh air. At first, people in the street thought her brother was her child! That was embarrassing for the young teenager.

Edda's little sister Petra, now five, was quite delicate in a number of ways. As a small child, she was extremely sensitive to scolding or shouting and fainted easily. Even a disapproving look was enough to make her cry. She was easily scared of strangers, most men, and especially doctors. When she started school, Petra's learning difficulties were noticeable right away. Her comprehension was a little slow, and she would have difficulty keeping up her grades.

Though Edda was used to taking care of her sister and brother, they represented responsibilities to her more than playmates of any sort. Because she was so much older than her siblings, Edda tended to think of herself as an only child. She did not feel lonely, though, as her mother had growing up. Edda enjoyed the company of adults and always had lots of friends.

One Sunday, Edda had plans with a girlfriend and did not go, as usual, to a movie with her grandfather. When she got home, Max had moved out. Apparently, Max had felt lonely and a little depressed that day. He had called the midwife who had helped deliver Herbert and said he would take that room for rent in her house after all. He told Margarete that he thought that arrangement would be better for all involved. Edda felt responsible and a bit guilty when her mother told her.

Though their movie dates ended, Edda and her family visited Max at his new place on many Sunday afternoons. Max enjoyed entertaining. They sat together in the garden, drank lemonade, and sometimes ate supper together.[27]

Max slipped Edda some money from time to time for things like amusement rides at Kinderfest, a family celebration held each summer. School children participated in the festival's parade with floats and costumes designed by each class and their teachers. Community marching bands played, and kids competed in sports and performed traditional folk dances. Heinz dressed in *lederhosen* and served beer in a large beer tent each year.

Edda was fourteen when the family moved in 1956 to a bigger apartment nearby with an extra room, a separate kitchen, a bathroom with a tub, and even a small balcony. Like many families, they had a garden plot with a shed close by. They could grow vegetables and flowers and relax there in the evenings after supper and on the weekends, enjoying the outdoors and the sunshine all summer. Petra and Edda shared a small bedroom, and Heinz had an office with a telephone. A phone was a rare thing in Germany at that time, so friends and neighbors sometimes asked to use their phone.

Heinz had finally found a business that suited him well, selling lightning conductors. With the German economy on a growth curve and plenty of jobs, many people started building their own houses, and they wanted to protect their properties from lightning. Tall structures in particular needed lightning protection. Heinz was a good businessman. He had a natural way of talking to farmers, homeowners, businessmen, and government officials. His business grew, and he soon had several employees. The family now had financial stability. He traded the VW for a bigger car and began planning a house for the family, to be built on the outskirts of Göppingen as soon as he had enough money saved to finance it himself without accruing debt.

By this time, a new modern Lutheran church had been built in their community, and Edda's parents began attending regularly. Edda

---

27. Max rented a room from the midwife, who had outlived two husbands, for the last six years of his life. He kept working until he retired at age seventy-five and died just before Edda finished high school.

went every Sunday and took confirmation classes. Her confirmation ceremony was a big celebration. All her aunts, uncles, and cousins came.

The youth-oriented pastor of the church and his wife were a young modern couple. They put together activities for young people in the church group: plays, singing performances for seniors, collecting money for charities, and spectacular outings. Edda thought the pastor's wife was wonderful. She sometimes invited the girls for get-togethers at her home. She was "so cool." She wore white bobby socks with her skirts, like Americans. Over the next couple of years, they both took the group of girls on a week-long trip to Austria and a bike trip to Holland, which Edda found fantastic.

Her mother's vivid description of the great migration in the face of the Red Army in 1945 had resonated deeply throughout Edda's childhood and adolescence. Margarete felt the terrible loss of her native town, where she had been happy even in wartime. She spoke often and tenderly about her Silesian past. The loss of her homeland haunted her for the rest of her life. Wherever she lived, Margarete always felt like a refugee, an outsider. Edda felt her mother's loss poignantly, but had little understanding of the politics of the time. She saw the effects of war, the destruction in the cities, some in Würzburg and much more when she visited Heinz's uncle in Stuttgart. But all history lessons throughout Edda's school years had stopped before the war. She knew that Germany lost the war, that Germany was divided between East and West, and that the German national anthem had been discredited. Beyond that, she learned nothing specific in school, or at home.

Edda was aware that Margarete occasionally sent packages to a family who had moved from Festenberg to East Germany. Max had rented some rooms in his house to them during the war. But otherwise they had no relatives or close friends left in Silesia (now part of Poland) or East Germany. The young teenager had no direct connection with the divisions that resulted from World War II, or the larger forces at work in rebuilding Germany.

Edda was in high school before she learned anything about the Holocaust, and that was from outside sources, newspapers. She found what she read shocking, unimaginable. The entire subject remained murky for Edda because no one, ever, discussed it.

She had the impression that the German chancellor, Conrad Adenauer, was a good man everyone admired, and he fostered ties with the United States. She saw the US military presence continue to grow in West German society. The nonchalant American soldiers were de facto cultural ambassadors whose influence went beyond chewing gum, Coca-Cola, and rock and roll. Edda was well aware of the American soldiers, products, movies, music, and fashion that were becoming increasingly familiar to everyone in Germany. The US Army base in Göppingen hosted friendship days to allow the local population and the soldiers to interact. The Blue Angels appeared fairly often, performing in exciting air shows that demonstrated flying skills with stunts, maneuvers, and special formations.

Edda never had any personal contact with any American during those early years, but she saw soldiers around town sometimes driving big American cars. "That was quite something to see." She and her friends sometimes giggled amongst themselves when they saw American women wearing bobby socks, rather than stockings, in winter. The girls wondered, "Don't they freeze?"

Overall, the relationships between the US military folks and the townspeople were friendly. There were certain areas, some bars especially, where soldiers congregated. Girls like Edda didn't want their reputations to be compromised by being seen in those rougher places where the military police were on call due to occasional bar fights.

The American presence included many black soldiers. These soldiers with African roots, but with American speech and demeanor, did not experience quite the same level of racism in Germany that they knew at home in the States. There were a few racially mixed children from the war in Edda's grade school. Edda thought they looked so exotic. She wished she had a black doll.

Edda was increasingly fascinated by the changing cultural landscape. Some family friends who had a relative in the US received a package one day that had a dress in it in Edda's size, and they gave it to her. It was beautiful and "very American." The summer dress had bold yellow stripes and a form-fitting elastic waist. The shoulders were accentuated with wing-like short sleeves. Edda felt so special just because the dress was from America. Everything American was intriguing to her. She studied English for three years in high school and loved the language. She also studied French, but never liked it. Her reading, writing, grammar, and spelling in English were good. Speaking English was not stressed in school, but Edda spoke English with her friend Gisela sometimes just for the fun of it.

Edda sang along to American songs: "Michael, Row the Boat Ashore" and "Ninety-nine Bottles of Beer on the Wall." Edda wasn't alone in her love for America. Oma Emma, who still visited regularly, loved Doris Day and her song "Que Sera Sera." Edda was a fan of Perry Como but thought Elvis Presley was "the ultimate," and often sang the words to "Love Me Tender," her favorite song. Edda saw many American movies, including *Porgy and Bess* and every Elvis Presley movie.

After Edda finished school at the end of eighth grade, she spent three years in a *Handelsschule*, similar to junior college business training. The emphasis was on languages and advanced math subjects.

When Edda was sixteen, she had a boyfriend who had adopted the name Dan to sound American. Dan was tall and artistic and tuned in to the American scene. He and Edda loved the same music and movies and took long walks together in the woods. Dan was a good kid, an only child who had a close relationship to his grandmother. He had a sweet and engaging personality and always thought of interesting things to do. They saw each other in school, after school, and sometimes before school when Edda would tell her mother that she had an early class. Edda was crazy about Dan.

The two spent all day one Sunday on a particularly long bike trip and came back late. It was dark, and her parents were worried. When she arrived home, her father was furious. Heinz called Dan's parents and said the two couldn't see each other anymore. Both sets of parents agreed it was over between Edda and Dan. Edda resented her parents for interfering, but she obeyed them. She still saw Dan in school since they were in the same class, but they were no longer friends. Dan moved on eventually and dated another girl at school.

The following summer, Edda met one of the young recruits at the National Guard base located nearby. She went out a few times with him. He was good-looking, slim, clean-shaven, and handsome in his uniform. They went swimming together and to an occasional movie.

But Margarete told Edda that a twenty-one-year-old was too old for her. Margarete knew, from her own experience, the downside of getting tied down to one boy when so young. Again Edda listened to her mother and told the young man that she couldn't see him anymore. He still tried to make contact with her a few more times, but she did as her mother wished and never saw him again.

Another friendship was terminated by her parents while Edda was in her final year of *Handelsschule*. Edda and her girlfriend Karin skipped school one afternoon in February and hitchhiked to Stuttgart to see a *Carnival* parade. They had a great time on that adventure, but their teacher found out and called their parents, who decided they two girls were a bad influence on each other.

Handelsschule had prepared Edda for an office job in bookkeeping and finance, which she did for a year in a Göppingen company that ran a chain of groceries. The job didn't suit Edda at all. At nineteen, she decided she didn't want to sit in an office anymore. Further education required going to Stuttgart every day, and that was not an option. But through her Lutheran church group, she was able to work in a one-year apprentice position in a church-affiliated hospice in the town of Wildbad in the Black Forest.

This was a sweet arrangement. The hospice was a spa-like hotel where people went for various treatments, therapies, massage, and special diets. The entire town, like Baden Baden, was built around healing water treatments and rehabilitation for those recovering from surgery or an injury, or who had chronic illnesses. The Black Forest surroundings were absolutely beautiful.

Edda's work involved kitchen duty and serving in the dining room, alongside other girls doing the same kinds of work. A youth leader coordinated activities for women apprentices to enjoy when they weren't working: handicrafts, bike tours, concerts, exploring nearby towns, and swimming in summer. This combination of work, leisure, and fun was done in a protected environment with a curfew and a ten o'clock bedtime. Edda was paid fifty marks (about twelve dollars) a month, as spending money.

The people receiving the treatments had fun, too, during the six-to-eight-week rehabilitation that was typical there. Edda loved to see people getting better. She learned a lot and had a good time. She enjoyed a short romance with her roommate's handsome brother, who visited Wildbad for a dance, but nothing developed beyond their brief flirtation.

With her apprenticeship year coming to an end, Edda, now twenty, was faced with a dilemma. After being more or less on her own for that year in the Black Forest, she could not imagine going home to live with her parents or working in an office again. She enjoyed her independence and loved the outdoor activities. Her idea of a "regular" German life felt very restrictive, without freedom or mobility. She didn't want to settle down in a small town, meet a boy, date, get married, and live close to her parents. The very idea made her feel cramped and tied down. She had always been considered a good girl who listened to her parents and always did what they expected. Now she was more than ready to break out of that mold.

Just before leaving the Black Forest, she happened to see notices in the ad section of a little newspaper that was published by the Lutheran church. Several agencies and families were looking for au pairs to care for young children. Edda was certainly qualified for that type of role, but her real interest was spending time in a different country and learning English. She considered her options.

Switzerland was a popular choice for many young women, but that didn't sound foreign enough to Edda. France had no appeal to her. She thought about England, but she realized that where she really wanted to go, what she really loved, was America! America was enticing, exciting, and new—the music, Elvis, everything American. Going to the US felt right.

One personal ad, not from an agency, said an American couple in Boston was looking for an au pair for a two-year contract. That was it! It sounded perfect. Edda applied immediately, and she didn't even tell her parents. To her delight, her letter, sent to an address in Frankfurt, was answered right away! She was instructed to come to Frankfurt for an interview with representatives for the family in America.

Heinz and Margarete were surprised when Edda told them about the interview, but they gave her permission to go. She took the train to Frankfurt for the meeting at a huge private estate belonging to a very wealthy couple. The people she met with there were in the publishing business and somehow connected to the American family. They passed along Edda's application letter, adding their own recommendation, to the family in America. Edda was told only that the family had two small children, ages three and one, and that they lived in the Boston area.

Soon a letter arrived from a Mrs. Ann Kissinger, written in German, saying that she and her husband Henry would welcome Edda to their family. The contract would be for two years and they would be her sponsors for obtaining an American visa. The Kissingers would also pay for Edda's boat passage, which she could pay back in monthly installments.

This was exciting news for Edda. But when she showed the letter to her parents, they were not at all pleased. However, they did not try to stop her. Edda was grateful for that because their disapproval might have swayed her. She said, "It was nice of them to let me go."

Margarete and Heinz realized that, at twenty, Edda wanted to see the world. They knew that a lot of young people did this kind of thing, and they figured she would surely return to them after two years.

Edda had an appointment at the American embassy in Munich to get her visa. The long process required a lot of health tests, a complete checkup. "They were very concerned about tuberculosis in particular." Edda spent an entire day there, with an interview, too. Edda thought they were very thorough; "They wanted to make sure."

In the American Embassy cafeteria, she had an American hamburger, and also ketchup, for the first time. On the menu it was listed as American Beefsteak. "It was very good!"

Even better, by the end of the day Edda was delighted to receive her immigrant visa.

She had stayed with some friends of her parents the night before, and she stayed that night again with them in Munich—nice people she'd never met before. When she went home to put her things in order and make travel arrangements, she found she had received another letter with details about her arrival in New York. Mrs. Kissinger's mother, who lived there, would meet Edda at the boat dock, then put her on a train to Boston.

The least-expensive ticket to the US was a nine-day journey by boat, a German boat called the SS *Bremen*, out of Bremerhaven, a city north of Bremen. She had her train ticket, boat reservation, and her American visa. But her departure was not a celebration for her family.

Her leaving was a blow to Margarete. Edda's mother added this loss to all the other sad memories she couldn't seem to get over. Petra was twelve, and "she cried easily and often anyway." Little Herbert was only six, and looked up to his big sister. He too cried when Edda left.

Heinz drove Edda, with her single suitcase, to the Göppingen station to board the night train to Hamburg. He was serious and sad as he hugged his daughter good-bye.

In spite of the melancholy expressed by her family as she left, Edda was happy and excited to be on her way. The all-night train arrived in Hamburg, where Edda spent the day with a friend from her Black Forest hospice job. Her girlfriend showed her all around the big city of Hamburg before Edda took the bus to Bremerhaven.

On the huge SS *Bremen*, Edda shared a cabin with two roommates. One roommate, going to Philadelphia, was seasick in her bed the entire trip. Otherwise, the voyage was pleasant. Edda met a nice group of young people who were all venturing into the New World. The young men and women spent the nine days playing cards, dancing in the evenings, and talking about their dreams.

Even decades later, Edda had a hard time describing the feeling she had when the SS *Bremen* pulled into New York Harbor on the cloudy morning of June 9, 1962. The sight of the Statue of Liberty brought tears to Edda's eyes. "It was a very emotional moment." Though she had never been there before, she felt somehow as though she had come home.

After Edda passed through US Customs and Immigration in New York City, she was met by a lovely older woman. Ann Kissinger's widowed mother, Mrs. Fleischer, welcomed Edda in fluent German and escorted her to the train that would take her over the next several hours to Boston and on to the suburb of Belmont, where Ann would pick her up.

Edda's excitement overshadowed her travel fatigue as she came closer to her final destination that evening. But her enthusiasm turned to panic when she climbed off the train in Belmont and saw the large dog with Mrs. Kissinger. Edda took a step back and, for a few seconds, feared she'd made a big mistake.

"I thought I was going to die when I saw that dog!"

A smiling Ann Kissinger greeted Edda warmly in German and introduced her dog Georgie. When Edda saw how gentle and friendly Georgie was, her fear was quickly calmed. On the brief car ride, Edda noticed what an attractive woman her new employer was.

Upon arrival at the Kissingers' New England home, Edda felt a fresh apprehension. Everything was unfamiliar. Though the two-story colonial house was comfortable and not extravagant, it looked huge—"So many rooms!"—a large entrance hall, formal dining and living rooms, a den with a television, a laundry room, kitchen, and at least three bedrooms. A small room in the attic had been finished in preparation for their previous German au pair, a young woman named Heidi, who was going back to Germany now that her two-year contract had ended.

The children were already asleep, so meeting them and Dr. Kissinger would occur the next day. After an evening snack and being shown her accommodations, Edda was more than ready to go to bed. She felt uneasy at first being alone in her own room. She had never in her life slept by herself anywhere. "It felt so strange!" But exhaustion overtook her that night, and Edda fell right to sleep in her new American home.

CHAPTER 9

Small Gardens and Big Ideas

When Suri had arrived in Boston late on a Thursday morning in the summer of 1959, no one met his plane. In his naïveté, he had not written to anyone with his flight arrival time. He had only provided the date. In India at that time, there would only have been a single flight a day—not several, one after the other. He took a cab to Harvard and arrived in the early afternoon, bag in hand, and headed for the Harvard Botanical Museum where he knew Professor Manglesdorf's office was located.

The scene that Suri was walking into at Harvard was especially exciting, because the whole field of biology was in transition. Having already become famous for its strong emphasis on classical plant genetics and cytogenetics, Harvard was now at the forefront of evolutionary biology. The outstanding faculty at Harvard's Biological Laboratories, Bio Labs for short, was rapidly moving into the emerging field of molecular biology. They sought to explain biological processes by understanding the interactions between different systems within a cell and the regulation of cellular processes.

Whereas cytogenetics was concerned with the structure and function of the cell, especially the chromosomes, combining cytology (the study of cells) with genetics (the science of genes and inheritance) allowed for the study of model organisms such as fruit flies, fungi, and bacteria, which were much easier to work with than crops. Crops took a full growing season, whereas these models could be reproduced in the laboratory. Many generations could be studied very quickly. The same transition away from plants was happening at other universities as well.

Though Suri was fascinated by such daring intellectual speculation, he was still interested in classical genetics, because it had a direct relation to plant breeding and plant improvement. It was a practical field rather than theoretical. His intention was to go back to India and work in plant breeding. He knew there was more possibility of his getting a good position in an applied field, particularly due to the Rockefeller Foundation's interest in India.

Professor Mangelsdorf was firmly rooted in the classical genetics of corn. His office and laboratory facilities were located in the basement of the Botanical Museum, along with his two postdoctoral candidates and his graduate students, one of whom was now going to be Suri. The postdoctoral fellows were Walton C. Galinat and Y. C. Ting.

Suri walked into the Botanical Museum building, saying, "Here I am!"

He was directed to the basement level where he met Walton Galinat, who had been assigned to make sure Suri was settled at Harvard. The two men apologized to each other when Galinat told Suri that he waited most of the morning at the airport for him, but had finally given up and left.

While they chatted, Professor Mangelsdorf walked in on his way to his lab. This was Suri's first face-to-face meeting with the esteemed professor, who made him feel very welcome. They chatted a bit about Suri's travel. He was glad to learn that the graduate students and post-docs came together for afternoon tea around that time, 4:00 p.m., each day. Manglesdorf's wife Peg, who worked with him, joined the group in time for tea and introduced herself to Suri. Gratefully anticipating some sustenance after so many hours of travel, and not having eaten since dinner the night before, Suri was startled when Galinat turned on the hot water from the faucet in the nearby sink and plopped a teabag into a cup for Suri. Tea time in India was quite a bit more formal and always accompanied by food—at least sweets.

Suri was starving, but he didn't dare ask about food; he still felt bashful around older people. He choked down the terrible lukewarm tea, while Manglesdorf explained that he had made arrangements for Suri to stay in the dorm at the Bussey Institution in nearby Jamaica Plain.

Galinat offered to show Suri around the Bio Labs building in the same complex. Suri had never seen so many labs before in one place. Every floor of the building was full of activity with students and professors going up and down the stairs. After the tour, Galinat gave Suri a ride to Bussey to show him his accommodations.

The Bussey Institution, almost seven miles south of Cambridge in Jamaica Plain, held an important place in the history of American agriculture. It was originally an undergraduate school of agriculture and horticulture, endowed to Harvard by Benjamin Bussey in 1835. By the early 1900s, Bussey had become a graduate school in applied science and plant anatomy and most of the original 300 acres had become the site of the stunning Arnold Arboretum.

Harvard had made a name for itself in classical genetics through the legacy of famous professors, such as Edward Murray East, a plant geneticist, botanist, agronomist, and chemist who led pivotal early research in modern plant breeding. In 1908, East, who was at Bussey, and George Shull, working independently at the Carnegie Institution in Cold Spring Harbor, each published their observations about heterosis, or what would come to be known as hybrid vigor, in corn. When certain inbred lines were crossed, the resulting offspring exhibited enhanced properties such as greater yield and sturdiness. Both East and Shull are credited with this important discovery, which, as the basis of hybrid corn, would eventually revolutionize corn production around the world.

Harvard transferred the Bussey staff to its Cambridge campus in the 1940s and folded the Bussey Institution into the Institute for Research in Experimental and Applied Botany.[28] When Suri arrived there in 1959, the large Bussey estate and grounds included greenhouses, experimental fields, and living quarters for graduate students and fellows. Living near the fields was an advantage in the summer months. By fall each year, most students found more conveniently located living

---

28. http://arboretum.harvard.edu/wp-content/uploads/VI_BI_2012.pdf.

arrangements in shared apartments or houses in Cambridge, but they returned each summer to Bussey.

A house in one corner of an experimental field was where Galinat and his wife lived on the ground floor. A German student lived on the second floor. An apartment connected to the greenhouse building was home to a caretaker, an interesting man from Latvia whom Suri enjoyed talking with.

The dorm at Bussey, called the Bull Pen (for men only), had four beds, and a spacious second-floor kitchen. Suri slept on the lower half of a bunk bed with no one above him. Everyone living in the Bull Pen, other than Suri, was American. He found their company a huge advantage those first few months, especially in learning more about his new surroundings and introducing him to American ways: using a Laundromat, buying groceries, canned and frozen foods, making sandwiches, dating, etc. However, Suri really had no time for dating that first year. He was busy full time and on weekends with his studies and his work schedule.

About five acres of experimental farmland at Bussey were used for research by Mangelsdorf, his associates, and graduate students. Professor East and his students had used the same land in past years. Suri was allotted a space in this field for his work. In India, he had worked only with dried specimens rather than live crops. By arriving for the summer before the academic year started, he was able to learn the procedures involved in actually growing a crop: pollination, cross-fertilization, and so on.

Suri became acquainted with various research techniques in the field. He mastered the intricate, complex breeding histories of his strains, adding to the knowledge Mangelsdorf and other great seed geneticists had developed. He and the other Bussey residents planted their own vegetable garden as well and enjoyed delicious sweet corn.

Mangelsdorf's research had revolved around the origin and evolution of corn. In the late 1930s he had developed his Tripartite Theory, which described the evolutionary relationship between maize and two of its relatives, teosinte and *Tripsacum*. The hypothesis had three main points: (1) The ancestor of cultivated maize is not teosinte, as was commonly believed by other scientists, but rather an extinct form of wild pod corn,[29] (2) Maize's closest relative, teosinte, is not an ancestor but instead derives from the hybridization of maize and *Tripsacum*; and (3) many types of maize have acquired, via introgression, genetic material from teosinte or *Tripsacum* or both.

Mangelsdorf had postulated that maize's ancestral pod corn was also a type of popcorn. Pod corn, in which each individual kernel is encased in its own little husk, was thought to have been grown in ancient times. Buoyed by archaeological evidence such as old cobs left by indigenous people thousands of years ago that fit this pod-popcorn description, he attempted to reconstruct the ancient species.

Suri saw the results in the field. His pod-popcorn behaved just like a wild plant. He noted, "Pod corn has very few seeds, only three to five at the most; Mangelsdorf, although a skilled plant breeder, prided himself on developing the world's most unproductive corn! As his students, we received pretty good exposure to corn breeding while working with him."

Suri stayed in the Bull Pen at Bussey that first winter, mostly because it was free. But his daily commute was time-consuming, almost an hour, and sometimes very cold. He had a ten-minute walk to the Jamaica Plain subway station; twenty-five cents took him to Boston where he transferred to a subway train into Cambridge; then he had another ten-minute walk through Harvard Yard to the museum complex. Getting used to the university academic system took focused time and effort. Examinations in India had always been in essay form. Suri found multiple choice exams very strange.

Galinat and Ting were valuable colleagues who worked closely together with Suri in the same offices and drank tea together. Afternoon tea times were much more enjoyable once Suri bought a single cup plug-in device to boil water in his cubicle.

He ate his meals in the cafeteria where he met his first good friend, an American student named John Koob. John was a geology student in the last year of his undergraduate degree. Suri sometimes ate dinner with John and his girlfriend Teresa.

Suri's first visit to an American home was on Thanksgiving when

---

29. Now it is generally accepted that teosinte is the progenitor of corn and not wild pod corn.

Dr. and Mrs. Mangelsdorf invited his graduate students and fellows to their residence at the Ambassador Hotel, which was decorated with carved pumpkins and corn of different colors. Thanksgiving dinners together would be a regular event as long as Suri was at Harvard. Mangelsdorf did the carving of the turkey, and Galinat told the story of Thanksgiving and how the tradition began right there in New England. Suri found the meal delicious and the entire occasion delightful.

In December, John Koob stopped by Suri's lab and asked if he had plans for Christmas. Suri did not know that making plans for Christmas was an expectation. John invited him for a late afternoon dinner at his parents' home on Christmas day.

Suri agreed, "Yes, I will go! Thank you."

On Christmas Eve it snowed five inches. Suri gazed in wonder at all the white bushes and trees, having never watched snowfall before. "Everything looked so beautiful!"

John showed up around 1:00 p.m. on Christmas day to pick up Suri and immediately handed him a very nice box of chocolates. Suri was surprised and thankful, but unsure why he was given this gift. Before leaving with John, Suri remembered a little ivory necklace his mother gave him to use as a hostess gift if he was invited to someone's home. He quickly tucked it into his pocket. After the lovely Christmas dinner with John's warm and welcoming family, and just as he was about to leave, Suri said hesitantly to John's mother, "I have this small thing for you." He pulled out the unwrapped necklace from his pocket and handed it to her.

Mrs. Koob smiled happily and put on the necklace right away, thanking Suri profusely.

At the end of his first year at Harvard, Suri was awarded a $1,867 Anna C. Ames Memorial Scholarship due to his high grade point average. That was enough to cover the following year's tuition. In addition to his full-time summer job with Mangelsdorf, at $1.50 an hour, Suri worked part time that summer in the Botanical Museum, doing odd jobs, counter work, and handing out admission tokens for the stunning collection of realistic glass flowers.[30] Due to his hard work and his

---

30. "Glass Flowers," a beautiful collection of glass models depicting hundreds of plant species, was commissioned by the Harvard Botanical Museum and made in Germany by Leopold and Rudolf Blaschka. http://www.hmnh.harvard.edu/on_exhibit/the_glass_flowers.html.

willingness to take any job that was available, Suri didn't require any financial help from his family at any point during his years at Harvard.

His second summer at Harvard, Suri's social life branched out just a little. He and John Koob and two other friends went on a hiking trip into the mountains of New Hampshire and slept in a log cabin owned by John's parents. John and Teresa took Suri to a square-dance party where a beautiful young woman pulled him onto the dance floor. He told her he didn't know how to dance, and she chirped back happily, "I'll teach you!"

He was glad that "she took pity on me." But he never saw her again after that evening.

Suri made another good friend, a graduate student from Bogor, Indonesia, who worked with Richard Evans Schultes on the top floor in the same building. The Indonesian's first name was Wertit, but he said to call him by his nickname, Tit.

Dr. Schultes said quickly, "No no, we don't use that word as a name." So Tit went by his middle name, Soegeng (pronounced Soo-gung), from then on.

In the fall of 1960 Suri moved to Cambridge, renting a room with Soegeng on Trowbridge Street within walking distance of the Museum. Soegeng was on a Ford Foundation scholarship, so he had more pocket money. He was generous, often suggesting, "Let's get some lunch," and sometimes paying for it, knowing that Suri had little cash. They saw movies occasionally, including *Lawrence of Arabia* and *Breakfast at Tiffany's*, in a theater on Brattle Street close to Harvard Square.

Just before Christmas that year, Dr. Schultes and his wife invited a few students for a holiday dinner: Suri, Soegeng, Raju, and Carluccio. Mrs. Schultes' dilemma was deciding what to cook. Suri did not eat beef; Soegeng was Muslim and did not eat pork; Raju was vegetarian; and Carluccio was Catholic, and it was Friday. She knew Catholics didn't eat meat on Fridays. The meal she served was fine though, and everyone had a good time.

Suri especially enjoyed the company of Dr. Schultes, who was a lecturer in Economic Botany. Suri loved the course on Plants and Human Affairs taught by Mangelsdorf and Schultes. Schultes had spent time in developing countries and had explored the Amazon. He had an informal demeanor, which Suri appreciated, and told many fascinating stories about indigenous peoples. A popular speaker, Schultes was often invited to give evening presentations to garden societies and Rotary clubs. When he needed someone to help with projecting his slides, Suri became his willing assistant, allowing him to learn even more about the professor's remarkable experiences and at the same time earn a little extra money: ten dollars for the evening.

Suri finished his classes in the spring of 1961 and his focus shifted to research in preparation for writing his thesis. He spent many evenings in the library and another summer back at Bussey.

That fall, he worked as a teaching assistant for a class in genetics and biology. He rented the ground floor apartment of a house on Sacramento Street in Cambridge with a roommate, a civil engineer named Giovanni. Giovanni's brother was a mechanical engineer named Guiseppe, but he preferred being called Peppino. Peppino, who worked in New Hampshire, was a regular weekend visitor. The roommates were later joined by a graduate student in plant physiology from Mexico, named Josué, who shared a room with Suri.

Suri's social life expanded considerably living with these guys. Josué was a fantastic dancer, and Giovanni was an excellent cook. They went often to the International Center where they met lots of young women. Most of the foreign graduate students in those days were from Europe. The roommates were included in various special occasions, such as when a very wealthy older woman invited ten students to her beautiful estate in Waltham for a spectacular Christmas dinner.

On New Year's Eve, Giovanni invited Suri to go with him and his brother to a New Year's party. Though Suri didn't drink alcohol before that, he went along; and it was quite a party with "wild and beautiful" young women, dancing, and drinks. Suri gained some extra attention as the only stranger in the group. Three "absolutely gorgeous" Italian girls, named Lina, Tina, and Antonietta, were teaching him how to dance. As the party got louder, the neighbors called the police. Someone suggested that they just move the party. Everyone loaded into cars and continued to an apartment owned by a French fellow. The singing and dancing and drinking continued into the wee hours.

Suri woke up the next day with a hangover and a blazing headache. Though he didn't continue drinking much after that, he began dating a little. Suri and his roommates threw parties at their apartment and lots of students came, including the beautiful Italian women. He started dating Antonietta, a student at Boston College majoring in French. Their dates consisted mostly of lunch together every week or so through the spring and early summer. Dinner would have cost about five dollars each in those days, and Suri did not have the money for that. He was still working most evenings anyway.

Back at Bussey in the summer of 1962, Galinat invited Suri to live at no cost on the second floor of the house, now that the German student had moved out.

Suri's friend Giovanni continued to think of fun things for them to do together on weekends. He had a car and a steady girlfriend. He typically invited Suri and Peppino along on whatever they did.

One morning Giovanni telephoned Suri and asked, "Do you want to go with us to Cape Cod and spend the day at the beach?"

Suri said, "Sure!"

Giovanni's girlfriend Irmgard was sitting next to him in the front seat when he pulled up outside Suri's place at Bussey. Peppino was in the backseat next to "the most beautiful girl!"

CHAPTER 10

Exotics

On Edda's first morning in the Kissinger home, she awoke ready to begin her new job. She found both children adorable right away. Elizabeth, with curly dark hair, was three years old and beautiful like her mother. She was a bit cautious toward Edda at first. The little girl had become quite attached to the departing Heidi. She would take a few days to warm up to Edda. Edda found cute, blond, one-year-old David very sweet and easy to care for.

Edda would have one day off a week and every other weekend for the next two years. Her responsibilities were primarily related to childcare. She would accompany Ann and the children on various trips to visit relatives. They would stay with Ann's sister and her family in White Plains; with Henry Kissinger's younger brother, Walter, and his family, at their beach house on Long Island; and they would visit Ann's mother at her apartment in New York City. On those trips, Edda would have extra time off to explore the Big Apple. Edda helped out during one regular event held at the end of each summer, when Dr. Kissinger hosted a garden party for graduate students and foreign dignitaries who had participated in his summer seminar series at Harvard on international affairs.

In the coming days, Edda learned more about the Kissingers. Ann was a loving and affectionate mother and a kind woman who treated Edda like the newest member of the family. From the start, she and Edda were on the way to forming an enduring friendship.

Though Henry Kissinger was somewhat distant, he was always considerate and kind to Edda. She liked the way he was with his children, "very playful, not very serious or strict."

He traveled a lot, "always coming and going." When he was home, Edda usually saw him reading newspapers and journals.

The Kissingers were German Jews. The families of Henry Kissinger and Anneliese Fleischer had come from northern Bavaria, not far from where Edda's family relocated after the war. Both families left Germany in the 1930s and settled in the Washington Heights area of Upper Manhattan, where young Henry and Ann both grew up.

Many Jewish families fleeing Nazi persecution had ended up in that neighborhood, and that continued through the forties and fifties. With so many German immigrants in Washington Heights, New Yorkers nicknamed it the "Fourth Reich." For a while it was called "Frankfurt on the Hudson." Mrs. Fleischer and the Kissinger parents still lived in New York.

Ann Kissinger's father had loved Germany, where the family had been happy and successful. He had never gotten over what had happened in his home country, the shock of how Germany could have done what it did. Ann had been close to him. She told Edda, "He was a sensitive man, and the war experience hit him hard."

Henry Kissinger had a beautiful office and library over the garage, with ultra-modern Danish-style furniture, big glass windows, and bookshelves overflowing with books and periodicals. Based on the book titles, Edda could tell he was interested in political issues.

While dusting in the library, she looked more closely at all the framed photographs. When she saw pictures of Henry Kissinger with German Chancellor Adenhauer and other world figures, Edda thought, *He must be an important person!*

Their neighbors were also quite distinguished. Derek Bok taught in the Harvard Law School, and his Swedish wife, Sissela, was an academic philosopher and ethicist whose parents were both noted political and social scientists.[31] The Boks had two daughters, Hilary and Victoria,

---

31. Sissela Bok's parents, Gunnar and Alva Myrdal, would both become Nobel laureates. Alva had served as Sweden's ambassador to India from 1955 to 1961. Gunnar won a Nobel Prize in economics in 1974. Alva won the Nobel Peace Prize in 1982.

around the same age as the Kissinger children. They got together for play time at either house.

However, neither of the Kissingers nor their lovely neighbors behaved like people who considered themselves important. They were warm and friendly to Edda.

Ann Kissinger was almost motherly in her protective concern toward Edda. She worried that Edda might become lonely or homesick, and she planned to introduce her to some other Germans. She sent her to a square-dancing class at the YMCA and encouraged her to go to Harvard's International Student Center to meet people. Edda enrolled in an ESL (English as a second language) class in Cambridge, where she met a German girl named Irmgard and her boyfriend Giovanni. Edda was soon swept into a busy social life on her days off.

Barely three weeks into Edda's new job that summer, Irmgard invited Edda to come along to the beach with some friends. "We can swim in the ocean!"

Twenty-year-old Edda Jeglinsky's first impression that day of twenty-eight-year-old Suri Sehgal was vivid.

"Suri was the last one to be picked up. I didn't know what to expect. I had never known anyone from India. He looked so exotic standing there, waiting for us to arrive. I was fascinated. We drove up, and he smiled the most beautiful smile. I was quite taken aback."

As Suri climbed into the car and sat next to Edda, he had a fleeting thought that their meeting had perhaps been planned by his friend. For the rest of the day, Suri was aware only of this beautiful young German woman.

The five friends had a nice day at the beach. The Italian guys in particular were amazed that Edda and Irmgard went swimming in the ocean. The water was ice-cold, and their muscles stiffened up, so they didn't swim long.

Edda was wearing a bikini she had brought from Germany, which—she discovered by the responses of those around her—was fairly unheard of at that time. She was not trying to be risqué. After that day she bought a one-piece swimsuit and never wore the bikini again.

The group went out to dinner together that night before heading home, and they all went to the beach a few more times that summer.

After that, Suri and Edda got together for lunch or dinner fairly regularly, but not every week. Suri was tied to his research work at the university and wasn't always available on the one night a week or every other weekend that Edda had time off. Edda was enjoying life with the Kissinger family, and she continued to have fun with her group of girlfriends. She met other guys occasionally, when out with the girls, but nothing serious ever developed with anyone else.

In the fall of 1962 Suri moved back to the apartment on Sacramento Street in Cambridge, this time renting a single room upstairs from Giovanni and another Italian fellow, Nello. Suri had more flexibility with his time that academic year because he had passed his oral examination, and Professor Mangelsdorf had left for a sabbatical in England. Suri now spent most of his time analyzing data and writing his doctoral thesis. Edda sometimes visited his lab while he worked. She found his work interesting, and they discovered how similar their views were on so many important things. Suri found that she had a "precise" mind, which he noticed came in handy when she helped him organize photographs for his thesis.

As Christmas approached, Edda felt homesick for the first time. She became tearful talking about it with Ann Kissinger. Ann tried to comfort her and agreed to allow Edda to put up a small Christmas tree in her room. Edda decorated it with a few ornaments she had brought with her from Germany, some new ones from a Scandinavian store in Cambridge, and real candles. She was happy, and the kids loved coming to her room to see the tree.

On Christmas Eve, Edda felt even more homesick for her family. So Ann took Edda to midnight Mass in a little church outside Belmont.

"I was so grateful that she went with me herself."

When the Boks heard that Edda was homesick, they kindly invited her to spend Christmas day and evening with their family.

About that same time, Ann made a suggestion. The Kissingers had spoken German with Edda for her first six months, to make it easier for Edda. Now Ann said to her, "If we don't start speaking English, you will never learn it."

So they agreed to only communicate in English after that. Watching children's programs on television with the kids and reading children's books to them was helpful, too. Edda appreciated that little Elizabeth would correct her if she mispronounced a word.

It also helped that Edda then met a new friend, Pat, and her friend Liz who had arrived from Wales with her, and a few Irish girls. Pat worked in the same neighborhood, and she and Edda saw each other almost every day and became best friends. Together they frequented Irish dances, coffeehouses, and bars where live music was played. Edda loved the music. Most was folk music, and she started a record collection: Joan Baez, Pete Seeger, the Clancy Brothers. She especially liked Irish music, which reminded her of German folk songs and drinking songs. Edda was surprised at how much some of her friends liked to drink! Drinking alcohol was new to Edda, and she experimented some. Tom Collins was the popular drink, but Edda decided she preferred a sloe gin fizz. One night at a bar, she and her Welsh friends were excited to find Ray Charles playing the piano—in person!

Suri and Edda began to go out more often to dinners, movies at the Brattle Theater in Harvard Square, coffeehouses, and the International Club at Harvard for big parties and celebrations, such as Carnival. If Suri was busy, Edda went with her girlfriends. She remembers eating a lot of pizza in those days, "which was very popular and new for me," and frequenting Chinese and Greek restaurants. Though there was a German deli in Harvard Square, she found it far more interesting to eat new types of cuisine. Suri and Edda also enjoyed some great meals, good wine, conversation, and laughter in Giovanni's apartment downstairs. The Italian guys did a fair amount of cooking, and plenty of different people popped in and out. The sink was always full of dirty dishes, and empty Chianti bottles lined the top of the kitchen cabinets.

Ann Kissinger took a protective interest in Edda's personal life and offered advice from time to time. When a young man she met at the YMCA her first summer took Edda home, Ann cautioned her, "That boy is not for you." Edda listened and discontinued the relationship.

When Ann noticed that the relationship between Edda and Suri was becoming serious, she expressed her concerns. She urged Edda to think about the differences between them. Ann knew firsthand the

Peepal Tree

Suri's maternal grandfather, Rawel Singh Wadhawan

Edda's paternal grandparents, Emma and Gunther Jeglinsky, with baby Heinz

Left: Edda's maternal grandparents, Max and Ida Wiorkowski

Below: Edda's parents, Margarete and Heinz Jeglinsky, on their wedding day

*Right: Edda starting first grade
in Bavaria*

*Below: Edda at the beach,
Cape Cod, 1962*

Suri in Mussorrie, Western Himalayas, 1956

Edda and Suri, Harvard graduation, 1964

Bill Brown, Edda, and Henry A. Wallace in Jamaica, 1965

Suri and Edda, wedding at Brown's home, 1964

USS Boxer *aircraft carrier evacuating folks from Santo Domingo, Dominican Republic, 1965*

Transfer from USS Boxer *to small boats to Puerto Rico, 1965*

Suri with his family in India, 1965

Front row (L to R): Chander, Ranjiv, Rick, Ladi, Raman, Jay, Shashi. Middle row: Savitri, Brij Anand, Suri, Shabji, Shila, Kedar, Nirmal. Back row: Sanjogta, Rita, Ram Chatrath, Padma holding Sanjay, Wally Sabharawal, Santosh holding Sarat, Mitter Paul Mehra, Shakuntla, Karanjit Puri, Parsanta, Gita.

Margarete, Edda, Herbert, Petra, Heinz in Germany, 1965

Suri holding Kenny, 1966

Shahji and Shila, with Jay, Edda, and Ben

Edda in India, 1970

Oma Margarete and the boys, 1973

Edda and Vicki, 1977

Kenny, Vicki, Oliver, and Ben, 1979

Jay and Kenny in the basement in Urbandale

Suri's Pioneer photo, 1986

Edda, Suri, Alice and Bill Brown, Honolulu, 1981

Ben, Vicki, Kenny, and Oliver, 1989

Kenny, Vicki, Suri, Maureen, Ben, Edda, Oliver, and best man, Jay, 1994

Above left: Siblings: Kenny Ben, Vicki, Oliver, and Jay, Captiva, Florida, 2003
Above right: Ryan and Vicki

*Siblings: Parsanta, Padma, Suri, Kedar, Santosh, and Shakuntla in India, 2005
(Sadly, Sanjogta died in 1995, and Savitri died in 1998.)*

Sehgal Foundation, Platinum LEED Certified Building

Ben and Maureen, Ryan and Vicki, Veena and Jay, Oliver, and Kenny in Jamaica for Edda's 70th birthday, 2011

Edda and grandkids: Marcus, Jessica, Sabina (seated), Katarina, and Emmett, 2011

Suri

Suri with villager women, Mewat

Edda and Suri with breeders and Raman at Kanater Research Station, Egypt

Suri and Edda in a village in Mewat, 2008

Sehgal Family Reunion, Des Moines, Iowa, summer 2014

dangerous and destructive power of prejudice and bigotry. Edda listened to Ann and considered everything she said.

By the spring of 1963 Edda and Suri were a "steady" couple.

When Henry Kissinger went to Germany, almost a year into Edda's contract, she was touched that he went out of his way to meet her family. Heinz and her little brother, Herbert, now seven, met Dr. Kissinger in Stuttgart. He brought back a pair of lederhosen for little David. He told Edda that he enjoyed the visit with her father. He added, "You have a cute little brother, a typical German boy!"

Suri was finishing work on his doctoral thesis, "Effects of Teosinte and *Tripsacum* Introgression in Maize," when Professor Mangelsdorf returned from his sabbatical in England. Mangelsdorf read the thesis and commented favorably on it. But there were two or three more drafts to come before the writing in English matched the professor's expectations.

Suri's thesis topic had been selected from several suggested by Mangelsdorf because of its applications in plant breeding. Suri's research confirmed that the introgressed germplasm (transferred genetic material) in maize from teosinte creates variability and imparts hard structure to the corncob. Dr. Mangelsdorf used to say that teosinte is to maize what steel is to a skyscraper. This was significant because a strong solid cob is needed in machine harvesting. But the important discovery was that when strains that had been introgressed with different chromosomal segments from teosinte are crossed, they exhibit heterosis.

Suri submitted and defended his PhD thesis in June to a committee consisting of Dr. Mangelsdorf, a professor from Bio Labs, and one from the Herbaria. His doctoral dissertation was accepted and praised by the committee.

Chatting over tea soon after Suri's thesis defense, Mangelsdorf asked, "What are your plans now that you are done?"

Suri's only immediate plan was a summer job as a teaching assistant for a class in introductory genetics and botany. Mangelsdorf suggested that Suri consider gaining some practical experience in commercial breeding before returning to India, perhaps a six-month postdoctorate fellowship position. Student visas allowed eighteen months after graduation for practical training.

Mangelsdorf offered to write letters on Suri's behalf to the two best fellowship options: Pioneer Hi-Bred Corn Company in Des Moines, Iowa, and DeKalb Corn Company of Illinois. Both were highly successful seed companies, and Mangelsdorf knew William Lacy Brown at Pioneer and Sherret S. Chase at DeKalb.

Replies to both letters were immediate and positive. Mangelsdorf advised Suri that he would be better off going to Pioneer. He had high regard for his friend Bill Brown who was the assistant director of research there at that time. They'd first met when Brown was Mangelsdorf's student at the agricultural experiment station at Texas A&M. Mangelsdorf also knew the founder of Pioneer, Henry A. Wallace. During the time that Wallace was secretary of agriculture in the Franklin D. Roosevelt administration, he had shown a lot of interest in Mangelsdorf's work. After receiving a copy of Suri's thesis, at his farm in New York where he had retired, Wallace had written to Dr. Mangelsdorf, complimenting Suri's work and commenting that it "raised a host of questions."

Brown wanted Suri to come to Pioneer right away, but Suri had already accepted the summer teaching assistant job. So the plan was for him to move to Des Moines in September. The idea beyond that was for Suri to return to Cambridge after finishing at Pioneer, tie up loose ends, and attend his formal graduation ceremony in May 1964.

Since his PhD work was done, Suri did not go back to Bussey the summer of 1963. Instead, he and Soegeng moved into a spacious apartment on Memorial Drive by the Charles River in Cambridge. Suri and Edda's group of friends now included more couples.

Giovanni and Irmgard were married in June in a small ceremony late one afternoon in Cambridge. The friends all went out to dinner together afterward. Giovanni took a job in Portland, Maine, so their visits were less frequent. Peppino still returned every weekend from Manchester, New Hampshire. Now he was a permanent weekend houseguest at Suri and Seogeng's apartment. Suri and Edda and their friends went often to Cape Cod that summer, and stopped frequently at the International House of Pancakes on their way back.

The apartment was a true international community with all kinds of students meeting and mixing. Pretty Smith College students living across the hall stopped in a lot; one of them was interested in Peppino. The mood was always welcoming, the people always friendly and ready for a song, a drink, a meal, or a party.

The Manglesdorfs invited Suri and Edda to their home one afternoon in August for tea. Their casual pleasant time together left Edda impressed with the university, the scientists, and the work they were doing.

The date was approaching that Suri was leaving for Iowa, and neither Suri nor Edda looked forward to being separated. They agreed to stay in contact by letter and telephone.

In September Suri stored most of his belongings in a trunk at the Botanical Museum, bought a $200 car, and drove west.

Suri had a friend along on his drive to Iowa, an Indian student about to start a postdoctoral position in Boulder, Colorado. The night they arrived in Des Moines, they stayed at the YMCA for five dollars. His friend boarded a bus the next day for Boulder, and Suri headed to Pioneer.

Pioneer's corporate headquarters were in downtown Des Moines, and its research operations were in a small building in Johnston, a little town a few miles to the north. Suri was welcomed there warmly by his new colleague, Sam Goodsell, who had arranged Suri's temporary accommodations. Dr. Brown was due back from a conference in Europe the next week.

Suri found a permanent place to live in the nearby town of Grimes. The landlady, Mrs. Kuefner, was a widow. She charged Suri eight dollars a week for his room, and he would share her kitchen. She also allowed Suri to use her phone to stay in touch with Edda.

At his first meeting with Dr. Brown, Suri was given the option to choose the research he wanted to work on. Brown had a continuing project with corncobs. Combines were replacing corn pickers for maize harvesting, and the strength and structure of the cob was important. Suri decided to work on corncobs and also spend some time in the field with the breeders.

Brown took Suri under his wing, and the two developed a close relationship that transcended the workplace. In the first few months, they published two papers, one on cob morphology and its relation to combine harvesting in maize, and the other on the introgression of teosinte in Corn Belt maize. Their work on cobs attracted attention in the corn seed industry.

Suri developed a profound respect for Bill Brown, whose spiritual life contributed to his qualities as a scientist and his decency as a human being. He and his wife Alice were Quakers. Alice told Suri that Bill appreciated the sense of peace and quiet he found at Quaker meetings each Sunday.

When the Browns invited Suri to their home for Christmas dinner, Suri asked if he could bring a friend. They said, "Of course!"

The Kissingers had given Edda some extra time off for the holiday, and she was meeting Suri in Chicago a couple of days before Christmas. From there, they drove to the University of Notre Dame in Indiana to visit a friend of his, a professor of genetics, before driving to Des Moines the day before Christmas.

Mrs. Kuefner was the only person Suri had ever spoken to about Edda, and she had seen pictures. Mrs. Kuefner offered a spare room to Edda for the Christmas visit and took Suri and Edda to the nearby church on Christmas morning.

The Browns were delighted to meet Edda. They had not realized that Suri's "friend" was his girlfriend. They had assumed he was bringing one of his physician friends. Suri had connected with a number of other Indians in Des Moines, including three doctors doing residencies at Mercy Hospital. He had enjoyed many dinners together with them on weekends, mostly eating hospital food.

Christmas dinner was lovely at the Browns' beautiful home. Bill and Alice liked Edda instantly. Not only was she beautiful, she was obviously well mannered, poised, gracious, and smart. She made quite an impression.

After the holidays, Suri focused on finishing his projects. One evening in late March, he was working late in the laboratory when Dr. Brown stopped by to chat. The thought briefly crossed Suri's mind that his six months were over and perhaps Dr. Brown was coming to tell him it was time to pack up and leave, even though Suri's project was ongoing and he would need a few more months to finish it.

Instead, Dr. Brown asked Suri about his future plans. He told Suri that Pioneer decided to start a corn-breeding program in Jamaica for the lowland tropics, and he asked Suri, "Would you be interested in the job there?"

All Suri knew about Jamaica was that it was a popular vacation spot. Brown didn't offer further details about the job, and Suri didn't ask for any.

Dr. Brown said, "Think it over and let me know in a few days."

Suri called Edda right away to tell her about the offer—he already trusted her judgment in all important matters. The only thing Edda really knew about Jamaica at that time was that people in Germany sometimes put Jamaican rum in their tea! But Edda thought it sounded great to work for an American company in a foreign country like Jamaica. She told Suri to take the job.

Suri thought about it for a couple more days. He recognized the job in Jamaica as an opportunity to jump-start his career in an applied field rather than in academia. He knew an awakening of sorts was taking place in Indian agriculture. Newspapers were carrying stories about an agricultural revolution on the way. When he left India in 1959 there was high unemployment among educated people and literally no jobs. Now there would be opportunities if he returned with breeding experience. Having an independent breeding program was its own attraction. Suri would be working closely with Brown, and perhaps even with Henry A. Wallace because of his interest in tropical corn. Henry Wallace's brother, James, was the current president of Pioneer.

But the most important reason Suri decided to take the job was because of Edda. Though they had not yet discussed marriage, Suri felt that it was understood by then that they would eventually marry. He wasn't sure yet what Edda would do after her contract ended with the Kissingers in June but, by staying in the Western Hemisphere, at least Suri could easily stay in touch with her.

Suri told Dr. Brown that he was interested in the job. Since Suri didn't have any practical experience in breeding, he and Brown agreed that he would spend the summer in Johnston as a trainee in corn breeding before leaving for Jamaica in September. When Suri asked

Bill Brown if he knew of any job Edda could have if she came to Iowa for the summer, Brown immediately offered to hire her to organize his private scientific library. Edda accepted the offer.

Now, as Edda's two-year contract was coming to an end, she was shocked and saddened by the news that Henry and Ann Kissinger were getting a divorce. They had been married for fifteen years. "The kids were so young!"

Edda could see that the breakup was sad for Ann, and also Henry, but it did not appear to be especially bitter. The couple stayed on good terms and often spoke about the children. Henry continued to come to the house and take the kids for walks or to the park. Edda always went along. He sometimes stayed for dinner.

Ann relied on Edda for moral support and comfort after the divorce. She never spoke ill of Henry, and she asked Edda to keep anything she knew about their private lives to herself. Ann said she did not want to harm his professional reputation in any way.[32] Though Edda's job with the Kissingers ended, Ann and Edda had firmly established a continuing friendship.

By the time Edda arrived in Des Moines with all of her possessions, Suri was living in an apartment in Des Moines with a friend named Rao, a pharmaceutical representative he met through his doctor pals. Edda rented a room in a house a short distance from Suri's place. Her first purchase was a red bicycle. At a time when barely anyone rode a bike around town, she rode to work in Johnston every day on her gleaming red Schwinn. Everybody at Pioneer knew her as the bicycle girl.

Edda enrolled in two Spanish classes at Grand View College in Des Moines, and they both kept busy with friends and work. The summer

---

32. Edda took this request to heart. When Henry Kissinger became such a public figure, she decided not to tell anyone that she had stayed with the Kissingers, so that no one would ask probing questions. She never even mentioned it to her own children until many decades later. Ann Kissinger remained a close friend, affirming to Suri years later how special she thought Edda was, and reflecting about that time in an email to her, "You were a dear and kind support for me. Lizzie and David loved you, and my hard-to-please mother thought the world of you."

sped by quickly. As Suri's departure date approached, Edda wasn't sure what she wanted to do next. She had an immigrant visa, so there was no urgency. One option was to continue at Grand View, or she could enroll at Drake University to finish her college degree.

A month or so before Suri was to leave for Jamaica, a group of friends came over to Suri and Rao's apartment: an Indian couple, the doctors, and two nurses. When conversation turned to Suri's plans and what Edda and Suri should do, one of the nurses, Mary Ann, piped up with a smile, "Why not get married?"

The whole group immediately seized on the idea and started applying pressure in a lighthearted way. As Edda put it, "They ganged up on him! They were practically pushing him into it. Mary Ann even gave him explicit instructions: 'You have to bend down on one knee, ask for her hand . . .' Right there with me in the room! It was all in good fun, but it also made a lot of sense. He proposed that same afternoon, in front of everybody. And I said yes."

Later, Edda suggested that Suri write to her parents and seek their consent. He wrote the letter in English and sent it by airmail, which had a turnaround time of about ten days. Heinz and Margarete responded promptly, giving their approval to the young couple and adding a personal note for each of them. Edda's father gave Suri a bit of advice. He wrote, "Marry Edda for the qualities you appreciate in her, and always remember to value those qualities no matter where you live or what job you have."

Edda's mother struck a different tone. Margarete wrote, "All my dreams of you coming back are now gone. I am over here, far away, and I won't have my Edda around anymore."

For her it was now final. Edda was not returning to Germany.

Edda said, "She eventually accepted it and moved on, but it was not the last time she made me feel guilty."

Suri decided not to tell his parents just yet that he was getting married.

The Browns were very happy for the couple. They offered their home for the ceremony and reception. The private ceremony was performed on the afternoon of September 26, 1964, by the Quaker minister. There were about fifty people at the reception. Dr. Brown had invited pretty much the whole department where Suri worked, anybody he had contact with at Pioneer, plus the handful of friends. Professor Mangelsdorf called during the reception so that he and the Harvard students could wish them well. Suri recalled, "The phone company had a slogan back then, 'Long distance is the next best thing to being there,' which was true!"

In the evening, after the reception, the Browns took Suri and Edda out for dinner. The newlyweds left for Jamaica the next morning.

CHAPTER 11

Land of Wood and Water[33]

When Bill Brown hired Suri Sehgal for the job in Jamaica, tropical corn breeding was new to Pioneer. Suri had to start from scratch, setting up and running the new research station. He soon learned that the initiative was not only supported by Brown, the project was of special interest to Pioneer founder, Henry A. Wallace, now in declining health due to amyotrophic lateral sclerosis (ALS, also known as Lou Gehrig's disease).

Suri had great admiration for the altruistic Wallace, whose influence was continuing to make a difference in agriculture worldwide. While vice president under Franklin Roosevelt in 1940, Wallace was deeply affected by the poor yields and primitive conditions of the Mexican corn farmers he saw during a drive from Laredo, Texas, to Mexico City to represent the US at the inauguration of the new president of Mexico. In long discussions, he and the American ambassador to Mexico, Josephus Daniels, concluded that the US needed to do something on a grand scientific scale to help modernize Mexican agriculture. Upon his return to Washington, Wallace found a receptive ear in Raymond Fosdick, the president of the Rockefeller Foundation.

The Foundation sent a fact-finding team, Paul Mangelsdorf and two other leading agriculturists, to speak directly with Mexican peasant

---

33. Jamaica's indigenous people called their island *Xaymaca*, meaning "land of wood and water" or "land of springs."

farmers. The Cooperative Mexican Agricultural Program was formed between the Foundation and the Mexican government in 1943. Crop scientist Norman Borlaug led the tremendously successful wheat-breeding research, which, in turn, prompted the Foundation to look at South Asia and the Far East. A committee, again including Mangelsdorf, traveled to Asia in 1951, which led to an agreement between the government of India and the Rockefeller Foundation in 1956 to modernize agricultural research in India, and the establishment of the International Rice Research Institute (IRRI) in the Philippines in 1960. The resulting rice-breeding program and the release of high-yielding rice from IRRI revolutionized rice production throughout Asia.

When Suri left India in 1959, much of the country suffered near-famine conditions due to the failure of the monsoon for several years in a row that resulted in severe drought conditions. Though the US supplied millions of tons of grains to India under the Food for Peace Program, more than humanitarian aid was needed to address the long-term food situation. Increasing the production of staple crops like wheat and rice was again seen as a viable solution.

A sample of the wheat varieties Borlaug bred in Mexico was planted in an observation plot at the Indian Agricultural Research Institute in Delhi in 1963. The results were impressive. When the rice developed at IRRI was introduced into India, it too was significantly superior. The high-yielding varieties of wheat from Mexico and rice from the Philippines provided the engine for growth that culminated in India's Green Revolution.

Suri's father had written to him frequently as he observed the agricultural changes occurring in India. Shahji was aware that India's agricultural system was being revamped with assistance from the Ford and Rockefeller Foundations and USAID. Land-grant colleges were being patterned after similar colleges in the US, which was bolstering agricultural education and research in the country and providing extension services to farmers.

Shahji knew there were many opportunities for Suri to obtain a good job in India. He made a forceful argument that Suri was now highly educated and needed by his family, and needed by India. He wanted Suri to come home, get married, and contribute to India's agricultural development.

Suri, however, wasn't ready to tell his parents that he no longer intended to move back to India. He was happy working at Pioneer. He was inspired and excited by the work to be done in Jamaica. And Suri was already married, very happily married, to Edda—something else he dreaded telling his parents. In their minds, marrying a foreigner was unthinkable. Both issues were highly charged.

When Suri and Edda arrived in Kingston, Jamaica, in 1964, they stayed for the first week in the quaint British-style Mona Hotel near the University of the West Indies (UWI). The tiny British car they rented seemed toylike compared to American cars in the sixties, and driving on the left took getting used to. The roads were pretty rough even when they weren't flooded; it was still hurricane season, and downpours occurred daily.

The small bungalow they found to rent on the outskirts of town was surrounded by lush, tropical vegetation, including a mango tree and bananas. A portable window fan was moved from room to room in the heat and humidity. The television broadcast only one channel, with a local program in the evening, a few British and American shows, and limited news programming. Somewhat isolated from the rest of the world while in Jamaica, Suri and Edda would have little direct exposure to the turmoil of the civil rights movement or the war in Vietnam, or other social changes going on back in the US over the next few years.

Edda settled happily into island life. She bought an old treadle sewing machine, made curtains, and focused on learning to cook. She had never attempted that skill before, but she dove in with the assurance that at least she knew how things were supposed to taste.

The first time she went grocery shopping, Edda rode to the supermarket on her red bicycle brought along from Iowa, causing an unexpected response from the people she passed. Young men and boys whistled and called after her; white women in Jamaica didn't ride bicycles in those days. She sadly sold her beautiful red Schwinn to an English woman at the university, who could ride it on the campus without causing a stir. Another problem was her reaction to mosquito bites. Edda was bitten so much that first year, she looked as if she had the measles.

Suri was receiving a salary of $600 a month when he started with Pioneer, which was later raised to $1,000—not a lot of money for a PhD in the US, but in Jamaica it went a long way. The only other work benefit was a company car.

A garden boy came once a week to take care of the hedges, grass, and plants, all with one tool: a machete. Since there was no washing machine, Edda hired a woman to do laundry and ironing. Rosita worked one day a week at first and later two days. She had two children, a lovely personality, and she was honest, unlike a maid Edda hired for a day and later found that things were missing. The only other times pilfering was a problem was when Margarete sent packages filled with goodies from Germany. Things tended to "disappear" at the post office or the customs office, and Suri experienced theft occasionally at the farm. Suri and Edda thought those sorts of things were understandable in a place where so many people had so little.

The majority of the impoverished population, which was about 1,740,000 people at that time, was made up of traditional peasant farmers. Jamaica had become independent only two years before, in 1962, after three centuries as a British colony. Jamaica's new national motto was "Out of many, one people." At least 90 percent of the population was of African descent; the rest were East Indians, Chinese, Syrian, and many combinations of intermarriages with Europeans: British, Swiss, German, and others.

Agriculture was a big part of Jamaica's economy, including sugarcane, bananas, and a modest coffee export market. Jamaica's tourist industry was still limited to the wealthy.

During the week, Suri was busy in the field, but on weekends the newlyweds explored the island and visited the pristine beaches on the north coast. Higher up the Blue Mountains, Suri and Edda sometimes stopped for a picnic among the pine trees where it was cool and the view was stunning. Colorful little houses dotted the Jamaican hillsides, roadside stands sold tropical fruits, and schoolchildren walked along in their uniforms. On Sundays, Edda loved to see beautifully dressed Jamaicans walking to church.

A main road wound around the island, and two crossed it. Smaller roads led up into the hills. People had warned, "Don't go up there! That's where Maroons live."

The Maroons up in the hills were the descendents of escaped African slaves who settled in Jamaica's rugged interior in the late 1600s. However, Suri and Edda never met danger from these people or felt unsafe on their sojourns through that beautiful landscape. They readily picked up hitchhikers and interacted with them to learn more about life in the countryside. They once gave an injured man and his companions a ride down to the hospital on the coast. The roads were in poor shape, so the trip took a couple of hours. These people were gracious and grateful.

Edda and Suri became friends with neighbors, folks from the farm where Suri worked, and with faculty at the university. A friendship with an English lecturer, who taught a botany course on evolutionary biology, led Suri to guest lecture on plant breeding and genetics in general. Suri appreciated the intellectual contact with students and faculty. Associations with UWI also meant regular social gatherings, holiday parties, and picnics, but at a much more laid-back pace than Suri had experienced at Harvard. Everything in tropical Jamaica was done at a slower pace. When Edda and Suri went to the community theater to see a popular political satire, *Eight O'Clock Jamaica Time*, not surprisingly the curtain rose at 8:30.

Caymanas, where Suri conducted his field research, was an old sugar estate in the Spanish Town area where the old capital had once been. The Caymanas estate had ten thousand acres of land mostly planted in sugarcane, bananas, and coconuts.

Suri would plant two crops each year in Jamaica, and Bill Brown would visit twice a year. Other folks from Pioneer and researchers from other breeding stations visited often. A Pioneer employee from the Canadian subsidiary visited for a few days to help Suri get started.

Suri hired all the labor for the nursery locally. There was no shortage of workers. The average daily wage for farm labor in the area was about ten shillings ($1.50). Pioneer paid more, fourteen to twenty-one shillings, depending upon skill level.

When Suri began the project, he worked out of a little wooden shack at the edge of a field. He and his team planted the first nursery by the third week of October, following procedures he learned at Jamaica Plain with Mangelsdorf and at Johnston with Pioneer. His job was to sort out all the germplasm (living plant seeds) from the diverse materials available from Brown's collection, make intelligent guesses about what could combine well with what, and then try to combine them.

Suri spent a lot of time in the field, studying the materials and preparing for the coming summer, the main crop-growing season. During the first winter season, he made many crosses intuitively from what he observed in the field.

In January 1965 Henry A. Wallace, his wife Ilo, and Bill Brown visited Suri's operation in Jamaica for a few days. The Wallaces had just been visiting corn-breeding stations in the Dominican Republic and Guatemala. They stayed at the Mona Hotel, and Brown stayed with Suri and Edda.

Meeting such a renowned figure as Henry Wallace was a great honor for Suri. He felt humbled to learn that Wallace already knew who he was when Brown inquired about hiring Suri for the Jamaica station. While at Harvard, Suri had been impressed by reading the "Century of the Common Man" speech Wallace made in 1942 that so passionately espoused freedom for all people around the world.[34] He thought Wallace was right on target with his revolutionary quote, "We hear a great deal about atomic energy. Yet I am convinced that historians will rank the harnessing of hybrid power as equally significant."[35]

Wallace and Brown spent most of a day with Suri in the nursery and walking the fields together, though Wallace was limping. Both men were enthusiastic and very pleased to see what Suri had done in the nursery, and the crosses he had made. The Wallaces came to Suri and Edda's home in the afternoon. Edda made a delicious apple strudel, which Wallace particularly appreciated as they enjoyed spirited conversation together.

---

34. Wallace gave his speech first in Spanish, then in English, at the Free World Association in New York City. http://www.youtube.com/watch?v=OBWula5GyAc. http://www.shafr.org/classroom_documents/Wallace,CommonMan.pdf.
35. *New Yorker*, Volume 60, Issues 16-24, 1955, 80.

Wallace gave Suri a lot of practical advice involving his operation, but he also wanted to accomplish several other things during his few days in Jamaica. Since he was breeding strawberries and gladioli at his farm in Westchester County, New York, he wanted to meet people who were working on those crops in Jamaica. Though no one in Jamaica was working on gladioli, Suri arranged a meeting with a scientist growing strawberries in the Blue Mountains. At Wallace's request, Suri also set up a meeting with the agricultural attaché at the American embassy and went along on the informative visit. The lively conversation focused on the application of genetics to improve crop plants. They then visited with the American ambassador, who briefed Wallace on the economic and political situation in Jamaica.

Wallace was passionate about plants and full of ideas. Suri found him to be a deep thinker and a genuine visionary who cared about making a difference in the world. Suri recalled something he'd been told by Simon (Si) Casady, a Pioneer associate back in Johnston who had been with the company from the beginning, serving as its first treasurer. In Pioneer's early years, Wallace had said, "Si, the guys running Pioneer think their job is to make profit. I suppose they are partly right. But that isn't our real job. Our real job is to learn how nature operates and to use that knowledge to make more food for the world."

Wallace's whirlwind visit kept Suri totally occupied. He confided to Suri at one point that he didn't have much time left, due to ALS, and there were so many things he still wanted to get done. Before leaving Jamaica, Wallace suggested that Suri expand his plant trials to the Dominican Republic, that it was an important agricultural country. He gave Suri the name of a Peace Corps volunteer there and encouraged Suri to contact him.

Suri took Wallace's suggestion. He found two pieces of land in the Dominican Republic, one near La Vega and another at the agricultural school in Santiago. First he went to plant the trials on five acres; then he returned a few times to check on the crop. For the late April harvest, Edda accompanied him to help Suri record grain moisture and take field notes.

Suri and Edda were busy harvesting the trials alongside other workers in the field near La Vega one afternoon when, suddenly, all the workers stopped what they were doing and crowded around a transistor radio.

Suri asked what they were hearing, and the answer was, "*Revolución en la capital!*"

Violence had erupted in Santo Domingo, a rebellion triggered by supporters of the ousted former president. The workers left the fields but Suri wasn't too concerned. The capital was about a hundred miles away and he wanted to finish harvesting the trials.

When Suri and Edda returned to the hotel in the evening, it was pretty much empty. A few employees stood around the TV in the hotel dining room. They advised Suri and Edda to leave, saying that almost every guest had already checked out. But not knowing where else to go at that hour, they decided to stay the night, with their door securely locked.

In the morning, when they found the hotel restaurant locked and the reception area empty, they felt a greater sense of urgency. Suri and Edda vacated their room in a hurry. But the battery was dead in their rental car. Luckily, the car was parked on a hill, so Suri was able to roll forward to start the engine.

They drove straight to the home of their Peace Corps contact, who accompanied them to the home of another associate, Raul Medina, the director of the tobacco research station in La Vega. Raul invited Suri and Edda to stay with his family. Friends visiting from Puerto Rico were guests there as well, and they were just as keen to get out of the country.

Each evening at the Medinas', everyone in the household listened to the *Voice of America* on the radio to follow the progress of what was now a bloody civil war. After three days, President Lyndon Johnson, fearing the Dominican Republic could become another Cuba in the Caribbean, sent in the Marines. All foreigners who wished to leave the country were told to travel to a corridor of safety around the Embajador Hotel in Santo Domingo for evacuation.

Suri and Edda decided to risk the drive to the capital. The Puerto Rican couple rode in the front car, and Suri and Edda followed. Both cars were stopped again and again by rebels pointing guns at them. The cars were allowed to proceed once the Puerto Rican couple explained in Spanish that they were all foreigners, and not affiliated with any political party.

Edda said, "The only thing they wanted from us was water."

Reaching the Embajador safely in the afternoon and showing their passports, Edda and Suri were directed through a barbed-wire barricade to the lobby, where US Marines registered everyone coming in. Within minutes, Suri and Edda were led to an open area within the hotel grounds to board a helicopter that flew them to the aircraft carrier nearby, the USS *Boxer*.

On the deck of the ship, Suri and Edda were impressed by the well-organized operation. They watched as helicopters kept bringing aboard more people, while fighter planes took off and landed, one after the other. The Marines were well-stocked with food enough for all the evacuees, and the soldiers gave up their bunk beds to civilians, an admirable gesture that was much appreciated.

After a few days, Suri and Edda and their friends were transferred to smaller boats for an all-night ride to Puerto Rico. Space on the boats was so tight that only the women were provided with bunk beds. Men had to sleep wherever they could find a spot. Suri was miserable and seasick the entire time.

Landing in San Juan in the morning, the evacuees were met immediately by the Red Cross, who attended to anyone needing help. The Puerto Rican couple kindly invited Suri and Edda to stay overnight in their home. Dr. Brown was immensely relieved to receive the phone call that Suri and Edda were flown safely back to Jamaica the next day.

The rental of land at Caymanas was for six-months at a time, an arrangement Suri found less than ideal. By the time Bill Brown came for his summer visit, Suri had talked to the Caymanas management and worked out a better deal for thirty acres of land on a long lease. After harvesting the corn crop, he wanted to plant legumes for crop rotation. He planned to plant red kidney beans, a frequent substitution for the popular pigeon peas that were not always available for the national dish: rice and peas.

The agricultural year in Jamaica was a continuous cycle of planting, pollinating, harvesting, studying the data, and planning the breeding nursery for the next season. Being busy year-round left little time for vacations. Suri and Edda had not been back to their respective homelands since arriving in the US, so they decided to take a trip between the August corn harvest and the first planting in late October 1965.

They flew back to Boston first and stayed with Peppino and his wife, Anna, for a few days. The couple had honeymooned in Jamaica only a few months earlier. Suri and Edda then drove to Providence to see Giovanni and Irmgard. The Mangelsdorfs arranged a small garden party with Suri's colleagues from Harvard.

From Boston, Suri and Edda flew to Germany. The plan was for Edda to stay three weeks with her family while Suri went to India on his own. Suri had still not told his parents that he and Edda were married, and he planned to tell them on this visit. Not sure how his family might react to his news, Suri and Edda decided that she would not join him in India at this time.

Suri was warmly received by Edda's family at their apartment in Göppingen, though he couldn't speak any German. Margarete was still unhappy that Edda was not returning to Germany to live, but Margarete and Heinz were pleased with Edda's choice of a husband. Petra was now a demure young lady, and Edda's little brother Herbert was an energetic kid. He and Suri had fun together with a lot of playful teasing.

During the four days before Suri left for India, the family took them sightseeing to Ulm, the *Bodensee* (Lake Constance), and the Black Forest. Heinz had brushed up on a few words of English, and Margarete spoke quite a bit of English. Suri appreciated their efforts, and Edda translated when language complicated communication.

Suri and Edda spent time in Göppingen and Stuttgart, buying electronic gadgets for Suri to take as gifts to India. With no imported goods available in India at that time, items from abroad were highly coveted.

When Suri left Germany in early September to fly to India, he had no idea that a series of border disputes and skirmishes, which had continued to occur with regularity between India and Pakistan since the Partition, had erupted into a war.

Landing in Delhi, Suri learned that his connecting flight to Amritsar had been cancelled. When he heard that the military had commandeered the civilian plane, he knew the circumstances were serious. The battles between India and Pakistan were initially contained within Kashmir, but the situation escalated when several other fronts were opened, including the Wagha border, about sixteen kilometers from Amritsar. As a border town, Amritsar was a likely target of attack. The only way for Suri to get to his family was by train or bus. He sent a message to his parents that he would arrive by train.

Despite of the potential danger, some of Suri's sisters and his brother Kedar came to meet his train that evening. Because Shahji knew so many people, both socially and politically, many other people came to the station to receive Suri as well. He was welcomed home with garlands of yellow marigolds, an Indian custom Suri didn't care for. He felt embarrassed to receive so much attention.

In the Sehgal home, relatives and friends, young and old, gathered around Suri. The younger children ran around, playing and shouting, "Uncle Suri is here!"

War raged for the entire three weeks of Suri's visit. He and his family remained pretty much housebound in light of the danger. Suri knew that Edda, in Germany, would be worried. The phones were not working, and the only place to mail her a letter was at the railway station. Though it was risky, Suri took off for the station on foot to mail his letter; Padma's husband, Ram Chatrath, went along. Suddenly there was heavy bombing close by, and planes circled overhead. Suri and his brother-in-law ran for shelter inside the railway station as they watched an airplane, hit by anti-aircraft gunfire, fall from the sky.

Meanwhile, Edda in Germany was more than worried. She was terrified. She had received a letter from Bill Brown inquiring about Suri's safety. Brown was very concerned, and so were many others back in Johnston. He had looked on the map and knew that bombs were being dropped right where Suri was. Edda and Brown were both relieved after Edda received Suri's letter and learned that he was okay. By September 22, India and Pakistan agreed to the UN-mandated cease-fire that ended the war. The war and Suri's visit with his family in Amritsar were both over.

In spite of all the drama and turmoil in the midst of a war, Suri had felt very well-received by his family. The most uncomfortable issue for him was that Shahji and Shila had wanted Suri to meet the young women they had selected for a possible match. He had not found the right opportunity to tell them he was already married. Suri wanted to tell them, but whenever he started to broach the subject, all his parents wanted to discuss was when he was going to return to get married.

To some extent, the reality of war rescued Suri from a true confrontation. There were no real chances to meet the potential mates his parents were considering for him. There had been too much danger.

No flights or trains for civilians were available when Suri left Amritsar, so he traveled by bus to Jalandhar to visit the family of a friend of his from the chemistry program at Harvard. Shahji and Shila happily joined Suri on the bus ride to spend more time with their son and allow Shahji to look into an opportunity he was considering at the time to establish a school for girls in Jalandhar. His parents continued traveling with Suri when he caught a train to Delhi, so they could be present for his departure. He managed to get on the plane without making any promises about his return to India.

On the way back to Jamaica, Suri stopped in the Philippines to visit people at the International Rice Research Institute in Los Baños, and in Mexico to pay a visit at the Rockefeller Foundation (IRRI), which by now was operating as CIMMYT, an acronym for *Centro Internacional de Mejoramiento de Maíz y Trigo* (International Maize and Wheat Improvement Center).

Once Edda and Suri were both safely home in Jamaica together, Suri wrote a letter to his parents, explaining that he was married. He inserted a photo of Edda that was taken on the beach in Cape Cod. The situation was final. Now his parents knew Suri wasn't coming back to India to be married, or to live. All correspondence from Shahji stopped completely.

Suri felt relief to have the truth out in the open. He continued to send regular letters to his father with updated news. Suri's brother Kedar now became the family interlocutor. He wrote that Shahji read Suri's letters to Shila, and that Shila sent her love to Suri through Kedar. He tried to assure Suri that their father's upset would be temporary.

After all the family issues, travel, and experiencing wars in two countries in the past eight months, Jamaica was a quiet atmosphere and pleasant safe haven. The place felt charmed to Suri and Edda. The Jamaican leadership was good, and the people were mellow and gentle.

There was little violence. The only danger Edda remembered was that too many people liked to drink rum and drive. "Driving could be dangerous, especially at night."

Edda and Suri loved the music: ska and now reggae. They went to concerts and listened on the radio to lively Jamaican bands, steel bands from Trinidad, and Harry Belafonte. Cricket was an obsession with Jamaicans. During matches that lasted for days, people were glued to their radios, including workers at their desks. Suri and Edda enjoyed watching the matches. Edda found it amusing that the players took official tea breaks, just walking off the field to drink tea.

Bill and Alice Brown and other folks from Pioneer were in Jamaica when Suri's red kidney beans were harvested in 1966. It turned out to be an excellent crop, and everyone was very impressed with what they saw. The beans were sold at a good price, and the income partially paid for the nursery. With two crop generations per year, within three years Suri would have excellent products that could be commercialized in the lowland tropics.

Each year Suri and Edda traveled to Miami to register, as required, with US immigration. They used those trips to shop for things not available in Jamaica, to go to the Fairchild Tropical Botanic Garden, and to visit the people at Pioneer's winter corn nursery in Homestead, just south of Miami.

When Edda found out, in the spring of 1966, that she was pregnant, she and Suri were elated. But for Edda, the thought of having her first child in Jamaica was scary. No one had ever explained childbirth to her. The natural delivery was not easy, because it was a breach birth. Kenai, later called Kenny, arrived a month early, in November. His name was a variation of the last name of a popular West Indian cricket player, Rohan Kanhai. Edda thought it was a pretty name.

The experienced Jamaican doctor was wonderful, and Kenny was healthy and fine. But the doctor kept mother and child in the hospital for two weeks so the baby could gain weight. Edda had a private room on the ground floor and participated freely with her infant's care. She could walk out onto a patio and in a garden, and at five each morning she was brought "bed tea," a British custom. With no family around to help her, Edda appreciated small luxuries.

Since it was near the end of the rainy season while Edda was in the hospital, there were still downpours day and night, and roads were flooded everywhere. Suri was expecting visitors from Pioneer, an entomologist and his wife, who were to stay in Jamaica for the winter. When they arrived, their wish to visit Edda was thwarted by the flooding. But Suri managed to come to the hospital almost every day. He felt a bit awkward at first with his newborn son, scared that the baby might break, but he soon got the hang of it.

Edda took the initiative with Suri's parents. She sent them a photograph of Kenny to introduce him to the family. She added the message, "This is your new grandson."

With the receipt of her letter, contact with Suri's family in India was restored. All heartache and misunderstandings were resolved and never referred to again.

Shortly after Kenny's birth, the family moved to Stony Hill. The house was bigger and a bit cooler than the old bungalow, with a beautiful view of the mountains. Rosita still helped twice a week. The only problem from Edda's perspective was the prevalence of cockroaches in Jamaica. When Edda got up during the night, usually twice, to get a bottle for Kenny, she switched on the kitchen light and saw huge black cockroaches everywhere! Edda screamed, and Suri would come to kill the scattering bugs.

The next summer, Suri planted cucumbers and made good money from them. Sweet corn generated some income, too. With the money coming in to support the breeding work, Suri needed very little funding from Des Moines.

The entire leadership of Pioneer came to Jamaica at one time or another while Suri and Edda were there, and they were very complimentary about what they saw. James Wallace and his wife visited Jamaica soon after his retirement from Pioneer. Sadly, his brother, Henry A. Wallace, had died only nine months after visiting Suri and Edda in 1965.

All guests visiting the Sehgals were invited to their home for a meal or for coffee and cake, and taken around the island to see the nursery and the plots at Caymanas. Many others came, too—some for short stays, and others, especially sorghum breeders, for longer stays ranging from two to three months. Bill Brown's twice-a-year visits, often with Alice, were the ones Suri and Edda looked forward to the most due to their warm friendship.

With a small child, Suri and Edda tended to have more contact with other people who had children. Kenny walked early and was very social. When guests came, he interacted with everyone and became sad when they left.

Edda decided to go to Germany for the birth of their second child in 1968, since there was no one in Jamaica to care for Kenny while Suri worked. Because the airlines had a restriction against travel beyond six months of pregnancy, Edda and eighteen-month-old Kenny left in April to stay for several months with Edda's family.

While Edda was gone, Suri's physician friend, "Pat" Pathak, stayed with Suri. Pat's family had gone to England for the summer holidays. When Bill Brown came to Jamaica for his usual summer visit, he stayed with the guys for a week. Pat was a good cook, although one evening he had a colorful accident when the pressure cooker exploded and sprayed kidney beans all over the ceiling.

Edda loved having time with her mother again. Her family adored Kenny, and everything about the visit was fun. Bernd[36] was born in July, and Edda and her sons stayed another six weeks in Germany. Traveling with two kids was not easy, and it was the first of many such trips to come.

Once back in Jamaica, Suri was happy to meet his new son, and Kenny started daily attendance at a Montessori preschool. Kenny loved going there. His teacher told Edda that, even though he was the youngest, he was the one who tried to comfort and console other children when they cried. The teacher thought Kenny was a good influence because other kids could see that he was happy there, so school must not be so bad.

Suri's work began to involve more travel. He went to Trinidad, Nicaragua, Venezuela, Haiti, and the Dominican Republic, often to conduct trials. He needed to develop different locations, at different times,

---

36. The German name Bernd, pronounced "Behrnt," is a shortened version of Bernard/Bernhard.

to measure how well his hybrids performed. All the steps from planting to harvest required several short trips of a few days each time. Edda didn't mind his being gone so much . . . except for the cockroaches.

Sugar estate owners were looking for other crops to grow besides sugarcane and bananas because international prices for these crops had dropped. Corn began to attract a lot of attention from the Jamaican government and the sugar and banana producers.

When Suri entered his hybrids in yield trials conducted by CIMMYT in Central America, they were the top three highest-yielding entries. Word got out to other parts of the Caribbean and Central America, and Suri started receiving visitors and requests from all those countries for seed samples. As the demand for hybrid seed increased, the need arose to establish facilities for seed production and conditioning. Suri built a small seed plant at Caymanas and began producing and selling to other countries in the region.

Suri's work in Jamaica was throwing him into the business world. The demand for his seed products forced him to learn on the job about production, sales, and distribution. His choices were carefully considered with a focus on common sense, and his results were profitable.

His success attracted a lot of attention in Jamaica and the neighboring islands. Each winter, Suri organized a field day at the estate, and invited the minister of Agriculture, Johnny Gyles, as well as other dignitaries from the Ministry of Agriculture. Suri's rapport with Gyles, visiting his office often to brief him and his chief technical officer on the progress of the breeding work, greatly aided Pioneer in obtaining concessions from the Jamaican government. They allowed Pioneer to import agricultural items for use at the station free of import tax. The minister eventually decided to replace the local corn variety, known as Jamaica Selected Yellow, with hybrid corn. Almost every week, there was news about hybrid corn in the local newspaper, the *Gleaner*.

All this activity in Jamaica started attracting serious attention from Pioneer's board of directors. Board members started taking an interest in what Suri was doing and what was happening generally in that part of the world.

When a plant pathologist named William Paddock convinced the Ford Motor Company, Standard Oil, and Pioneer to join hands in what turned out to be a short-lived project to "feed the world," Suri was asked to represent Pioneer in the effort to supply seeds for small farmers in developing countries. He traveled with the three-company team to Ecuador, Colombia, and the Dominican Republic.

After more than five years of Suri's demonstrated accomplishments in Jamaica, Bill Brown asked him in early 1970 about his interest in leaving Jamaica and possibly spearheading Pioneer's fledgling international operations. Suri was indeed interested.

After so long on the island, Suri and Edda had begun to feel isolated. Going to Miami every year with little children in tow was a hassle, and they were now anxious to move back to the US. Suri and Edda wanted to become American citizens, which required living in the US continuously for a minimum of thirty months. They were ready to go.

Bill and Alice Brown, accompanied by R. Wayne "Skid" Skidmore and his wife, visited Suri and Edda in Jamaica in early 1970. Brown was now vice president of research, and Skid was president and CEO of Pioneer. During the visit, Brown prompted Suri to tell Skid what he had done in breeding, production, and sales in the Caribbean and Central America, and to describe the opportunities he saw developing around the world.

Skid was sufficiently impressed by what he heard that he made Suri an offer on the spot to move back to Des Moines to build Pioneer's international business.

Suri accepted the offer, saying he would first like to have three months off to visit his family in India before starting his new job. He explained that he had just learned that his mother was ill with stomach cancer, and she wanted to meet her daughter-in-law and grandsons.

The men from Pioneer readily agreed.

With that, Edda and Suri sold or gave away their household items, and Suri shipped his books to Johnston. They left Jamaica in March 1970 with their two little boys and a couple of suitcases. And to their delight, Edda was newly pregnant!

CHAPTER 12

All Is Forgiven

Suri and Edda and their young sons flew directly to Germany. After a few days, Suri flew ahead to India. The plan was for Edda and the boys to have another week with Edda's parents before continuing on to India. But when Suri arrived in India, he learned that his mother's stomach cancer was advanced. Shila had been rushed to the hospital, and nobody could say how long she might live. Suri sent an urgent telegram to Edda, telling her to come right away. Edda answered his telegram immediately with the details of her travel arrangements.

Edda felt both anxious and excited to get to India and meet Suri's family. She and the children flew overnight in late March and landed in Delhi early in the morning. Before boarding the flight from Delhi to Amritsar, Edda took the boys to the restroom to change out of the warm clothes they wore leaving Germany, and to change Bernd's diaper. They had to step over the cleaning ladies sleeping on the floor, only to find that the toilets were squat toilets—a hole in the floor, footprints on either side, with no toilet paper.

As Edda changed Bernd's diaper near the sink on the floor outside, people crowded around to watch what she was doing. She would have to get used to being stared at in a place where there were so few white people.

A propeller plane took them to the tiny airport in Amritsar, which handled only a few flights a day. At the airport, the other passengers went their way, and the place was practically deserted except for Edda and the boys. Nobody had come to meet them; Edda's telegram had not reached Suri.

Edda knew the Sehgals' address, but there were four streets with the same name, listed as 1, 2, 3 and 4, and she did not know the street number. She tried to look up the telephone number in the phone directory, but there was no listing. The Sehgal phone was listed under a business name that Edda did not know.

The lone person on duty at the airport made a call to the airline office, and then sent Edda and the children on the airline bus to the city office. The roads were in terrible shape and Edda remembers, "That bus driver, wearing a turban, drove like the dickens! The boys and I were flying up and down on the backseat! My god, we almost hit the ceiling. It was quite a ride."

When they climbed off the bus in the city, a group of schoolchildren walking by in their uniforms all stopped, stared, and crowded around Edda and the little boys.

Inside the airline office, Edda and the boys sat down, tired, hungry, and thirsty after traveling all night. After waiting for about ten minutes for the man at the desk to look up, Edda stood up and asked him directly for help. When he saw the two little boys (Bernd was not yet two years old, and Kenny was three and a half), he began a series of phone calls to help find Suri and his family. The man had no luck until he called a banker friend and was given another number to try.

When the phone rang at the Sehgal home, Suri's youngest sister, Sanjogta, answered. The voice asked, "Is there anyone in the house married to a foreigner?" Sanjogta handed the phone to Suri.

The man on the phone said, "Your wife is sitting in the city office of the airline."

Suri asked, "Which city office?"

The reply was, "There's only one city office; you must know it!"

Luckily Parsanta's husband, Karanjit Puri, knew the city office was not far from the house. The two men took a rickshaw there.

Suri was overjoyed to be reunited with his wife and sons. However, they were unable to go back to the Sehgal home. The custom on auspicious occasions, such as a daughter-in-law's first visit, was for her not to enter the house without first having a religious ceremony, a *puja* (or *pooja*), a prayer or blessing by a *pundit* (Hindu preacher). In addition, the house was a bit messy with work going on to install a Western-style toilet.

Suri and Edda were advised to stay at the tiny hotel across the street from the hospital until arrangements could be made for the puja. After checking into the hotel, they were informed of another required custom: before meeting Shila at the old Victoria Jubilee Hospital, Edda had to first visit the Sikh Golden Temple to receive blessings.

Suri and Edda went along with these rituals without protest.

Edda and the tired little boys had been traveling all night and all morning, but they went to the Golden Temple. After receiving their blessings, they went straight to the hospital to see Shila.

A new hospital was under construction, but for now Shila was in the main ward of the run-down old hospital. To navigate the ward, visitors had to step over all the other patients' family members. A typical custom whenever someone was in the hospital, especially from Indian villages, was for the whole family to accompany the patient and sleep and eat on the ward, in the halls, and outside on the lawn. A white woman there with two small children meant lots more staring.

Shila, though quite ill, was overjoyed to see Edda and the children. She welcomed them with smiles and warm gestures, and blessed them each by touching the top of their heads. She held Edda's hand throughout the visit.

Edda could see how sincerely happy Shila was to meet her and the boys. Edda thought, *Once there are grandchildren, all is forgiven.*

During that first visit, flies were thick in the room. Edda was shocked when a worker walked up and down the ward with a Flit gun spraying DDT right over the heads of the wall-to-wall patients and visitors. The still-common practice filled the space with a nasty chemical smell.

Shila was not expected to live long, and all the relatives wanted to visit her and to meet Edda and the children. So the family arranged for the puja ceremony within a day.

Relatives gathered around Edda. They hugged and kissed her and the boys and gave her gifts. She felt overwhelmed by their sweet attention.

Edda was wearing a pink pantsuit that first day at the house. Before leaving Germany, she had bought some clothes that looked somewhat like Indian clothes. She knew skirts or dresses would not have been appropriate. In the Punjab, young women usually wore pants (*salwar*) with a long blouse. Edda's pink suit was just right. She later had a few Indian pants and tops tailored for her in Amritsar and wore them during the rest of her stay.

Suri recalled, "A beautiful woman in a beautiful pink suit; everybody fell in love with her."

The routine was now to visit Shila in the hospital each morning and then go back to the house, where at least fourteen children and twenty adults were present every day. Most of the relatives spoke English. Only Shahji and Shila did not, but that didn't matter because everyone helped to translate. Smiles and gestures were helpful, too.

Meals were at one big table in three shifts—first all the children, then all the men, and then all the women. Guests came often. The family had to send out for extra help. One helper came with a broom and swept the floors every day, and another came after every meal to do dishes. She brought her pretty young daughter who was about ten years old. The two of them did all the dishes every day and went on to do the same at other homes. Watching this, Edda yearned to take the sweet little girl to America.

When any new visitors arrived in the afternoon, someone in the household went to the corner store to buy Coca-Cola. Coke in a bottle was a novelty. Care was taken to assure guests that it was authentic Coca-Cola, so it was served in the bottle with a straw.

Afternoon tea was the best meal of the day, according to Edda. "Honestly, I loved it all. Shahji made sure that every day somebody bought fresh cookies from the bakery for our tea. That was so nice!" They had German gingerbread cookies, just like Christmas cookies. "I was in heaven."

All eyes were on Edda all the time, watching to see how she did things. Suri's family was used to thinking that Western women turned over their children's care to governesses. Before meeting Edda, they thought European women didn't really *do* anything except play bridge and rely on servants. But Edda helped in the kitchen, boiled water for drinking and cooking, did everything they were doing, and at the same time attentively cared for her kids. Kenny was always running all over the place with his cousins close in age. So Edda was always on the move, too. The kids had tremendous speed, and so did she, going up and down three or four floors, making sure no one fell off the roof.

Preparing meals was a big job with so many people, and with guests popping in without notice as well. Cooking went on all day on small kerosene stoves. Bread was baked in a *tandoor* (clay oven) in the courtyard. Edda took over the job of making salad for the whole family in the evenings, cutting up piles of cucumbers, tomatoes, and onions, and putting them all on a big tray on the table. Once in a while, she cooked something plain for her boys. She boiled potatoes or vegetables, or cooked eggs. Shahji especially enjoyed Edda's boiled cauliflower.

Suri was relieved that Edda integrated so well into the daily routine, because participation in the Punjabi household makes one a member of the family. Though nothing about the Indian diet was familiar to Edda, she ate what everyone ate: *chapatis* and vegetables for lunch, rice and dal for dinner. Fruit and yogurt were plentiful. Little Bernd didn't have any problem eating chapattis, rice, and dal, but Kenny had diarrhea most of the time. He finally refused to eat anything. He liked buffalo milk with Ovaltine. Though he did lose weight, he didn't appear to suffer from his limited diet. Kenny was happy as long as he could play. He loved bouncing and wrestling on mattresses in the courtyard with his cousin.

Kenny fussed whenever he had to go to any bathroom when he was not at the house, because they were "dirty and smelly." Edda was thankful that Bernd was still in diapers. She had used disposable diapers for him on the flights, but for the long stays in India and Germany, she had cloth diapers, which needed washing. Suri's sisters took turns doing the laundry by hand every morning. Edda and Suri's sister, Sanjogta, who was close in age and unmarried, became good friends. Some afternoons, they went to town by rickshaw for ice cream cones or a little shopping.

After a week, Suri and Edda and the kids moved from the hotel to the house. Shila's condition became more stable, and she was very glad to be able to move home with her family. The family doctor came every day to see her. Since she couldn't swallow food, she was fed through a tube inserted in her stomach. Suri took her for chemotherapy sessions. But Shila's condition was not improving, and Suri found it awful to watch his mother suffer so much. She would often plea that they not feed her. She always felt much worse after the feedings, and she did not want to be kept alive in this way. Edda, Suri, his brother Kedar, and his sister Sanjogta, who was Shila's main caregiver at the time, supported Shila's wishes in this regard. But Shahji and Suri's other sisters were quite emotional about the situation and still held out hope that Shila could be cured despite her dire condition.

Edda and Sanjogta spoke more than once with Shila's family doctor about discontinuing the feedings, which seemed like torture to them, and allowing Shila to have a dignified death. But the doctors insisted on keeping Shila alive as long as possible, saying that even prolonging her life for a few days was important. This situation was so disturbing to Suri and Edda that they decided to sign durable powers of attorney for themselves so this could never happen to either of them.

In the middle of their stay in India, Suri and Edda took the boys to Kashmir for a week. They stayed on a houseboat on Dal Lake and explored the breathtaking surroundings by *shikara*—a wooden boat similar to a Venetian gondola. From people in other small boats on the huge lake, they bought apples, cherries, and other treats. They then traveled by car and horseback to Pahalgam, a center of saffron production, where there was still snow in the mountains. Living in Jamaica, the boys had never seen snow, and they loved going uphill on horses and downhill on sleds.

Toward the end of their time in India, Suri and Edda visited some friends in Delhi. They all went to see the Taj Mahal, a few hours away in Agra. Edda was struck by the pure magic of the monument, the gorgeous grounds, and the serenity. There were no crowds, no guards, just unforgettable beauty.

The weather had been hot in April. May was even hotter, dry heat like an oven, with occasional dust storms. During the day, the electricity went off for hours, and there was no running water. Without water, they could not flush the new Western toilet. So before 9:00 a.m., the family filled up containers to have water for the rest of the day. The ceiling fans were useless without power. Edda noticed that, throughout the entire time she was in India, no one she was around complained about any inconvenience, no matter how big.

Before leaving Amritsar, Shila wanted a family picture taken on the patio. She sat with Shahji beside her in the center of the picture and the rest of the family all around her. Suri felt grateful for the peace and acceptance he now had again with his family, but he left India with terrible sadness, knowing it was the last time he would see his mother alive.

Suri, Edda, and the boys flew back to Germany where her parents now lived in a new, spacious house. Edda and the kids stayed there for a few more weeks, while Suri flew back to Des Moines to find a place to live.

That first journey to India, and the two-and-a-half months spent there, left a deep imprint on Edda. She had been a little anxious at first, but Suri's family turned out to be so wonderful to her. She experienced a kind of total love, warmth, and acceptance from them that she'd never felt or seen anywhere in the West.

Margarete noticed a change in Edda and asked, "What's the matter? You are so different."

But Edda could not answer her mother. She was lost in her own thoughts after her experiences in India.

Edda could not explain the way she felt having been so warmly received in India by Suri's family, and the circumstances that were so different from anything in her past. In Europe, Edda had been keenly aware of restrictions, obstacles, and judgments. In India, all the little details, rules, and expectations that seemed very important in the West didn't seem to matter. Anything in life could change in the blink of an eye. Yet despite any hardship or inconvenience, the people Edda met in India lived with joy and acceptance, without criticism or complaint— even Shila as she was suffering focused only on her love for her family. Edda's life with Suri in America had a sense of freedom that Edda appreciated now more than ever.

Her third child was not due until October, so staying in Germany for the birth was not even considered this time. Edda flew home to Iowa in the beginning of July. Word soon came that Shila had succumbed to the cancer, and Edda shared Suri's sorrow at the loss of his sweet mother.

By September 1970 Suri and Edda moved into a split-level house in Urbandale, just west of Des Moines, where the school district had a good reputation. The house had three bedrooms, an office for Suri, and an unfinished basement.

Edda's mother came right away to help out with the kids for a month before and a month after the birth of Oliver in October 1970. Oliver was another breach birth, but the entire delivery experience was a lot easier on Edda this time, and Margarete was a tremendous help. She dove right in, meeting the neighbors, making friends, sewing curtains, and keeping Kenny and Bernd entertained.

Edda's hands were kept full with three very active little boys. She developed friendships with other Pioneer families, especially those with small children. Having regular contact with Alice and Bill Brown again was wonderful, like having family nearby. The Browns were much like a local set of grandparents for the children.

As Edda held the home front together, Suri tackled understanding the dimensions of his new position. He brought work home in the evenings and on weekends. He flew down to Jamaica when needed to keep his eye on the research station.

The shape and scope of Suri's new job were his to determine, just as in Jamaica. He had already proven his skills as a natural entrepreneur, but his theater of operations had expanded considerably. Bill Brown's philosophy was to give the greatest freedom to his people and support them with adequate resources to assure success. His model was an ideal recipe for Suri in developing strong business relationships. Suri's career and Pioneer's international success were about to take off in a big way.

Edda juggled a lot of responsibility on her own as Suri was gone more often. Even when he wasn't traveling, he came home late in the evenings. He kept Edda apprised of each new development in his work, and she provided her wise counsel with any new dilemma.

Margarete came back again for an extended visit when Edda gave birth to their fourth child, born during a blizzard in April 1973. Edda was thrilled to have a girl this time.

The basement had been finished to include bedrooms for the older boys by then, all three of whom were balls of energy. Close in age, they got along well and played together. Edda and Suri involved their sons in the choice of a name for the baby. They agreed on Vicki. Bernd, almost five, was the most interested in the infant, but all three would give their sister little kisses.

Shortly before Vicki was born, three-year-old Oliver had been diagnosed with asthma. He was allergic to grass, weeds, pollen, all kinds of things. He would wake up gasping for breath. Edda and Suri spent time in hospitals with him frequently, because the medication available wasn't very effective. He was given a lot of cortisone, the only medicine that really helped. But asthma never slowed down Oliver. He remained a busy little guy.

Edda and Suri kept their kids well stocked with books. Kenny, especially, liked to read. When he was finally old enough to start kindergarten, after three years of preschool, he found it disappointing. He came home the first day and complained, "All we do is play!"

When he started grade school, he was placed in an open classroom and his teachers were frustrated. "Kenny doesn't do what he is supposed to do! He just goes and reads and reads."

Edda never tried to speak or teach German to the kids. Rather, she and Suri worked hard to keep improving their own use of English. Urbandale's ethnicity was pretty much all white except for one African-American family and two Indian families. Edda knew that the slightly darker complexion of her children could make them targets of bigotry. She wanted her kids to fit in and feel secure, wanted and loved, without emphasizing their Indian and/or German roots. Suri and Edda wanted their family to be American.

The kids would always remember the Fourth of July as the most celebrated holiday in their childhood, rivaled only by Christmas. The family joined in Urbandale's civic celebrations, such as kids' races in Lions Park, the parade, fire-hose battles between neighboring fire departments, and fireworks at night.

For Suri and to Edda, the holiday held more significance. Becoming US citizens was important to them—a strong desire, a lifetime goal. They had been counting the days before their applications for citizenship were granted in the spring of 1974. A ceremony was held in a downtown government building for the twenty-five or thirty people taking the oath of citizenship. Most were from Latin America; Suri was the only person from India.

Suri and Edda had enjoyed the informal privileges and freedoms associated with being immigrants for many years; but now that they were officially Americans, they were overflowing with feelings of gratitude. The first thing they did was to take the whole family to the passport office. Suri, Edda, and their kids had been carrying different types of passports whenever they traveled together. But now they each had a US passport. Suri and Edda were delighted beyond measure. They valued highly the freedom and opportunity they found in the US. Now they wanted to share those advantages with their families.

Suri did not see a bright future for his nephews and nieces in India. Employment opportunities were still extremely limited. He was determined to give his relatives the possibility of a better life in America or elsewhere in the world. He encouraged his brother and sisters to send their children to America for their education when the time came, saying that he and Edda were willing to help them in any way they could.

In the coming years, as Suri traveled all around the world, he tried to always be home for the important holidays. Independence Day remained the foremost in significance.

CHAPTER 13

Going International

Pioneer Hi-Bred Corn Company, though identified mostly with the Great Plains, was one of two premier seed companies in the US in 1970. With Suri in his new role, the company name was now changed to Pioneer Hi-Bred International, Inc. Building on his success in Jamaica, he would now take the company to countries around the world.

Working in the international arena meant navigating complex trade barriers and foreign exchange restrictions. Such impediments could be especially daunting in developing countries, as well as in the Soviet Bloc. Differences in business cultures had to be understood, worked through, or negotiated with every country. Success in international trade required an unusual flexibility in executive decision-making, including the ability to respond quickly and improvise short-term fixes when needed. Suri embraced the challenges with skill and thoughtfulness.

Adopting the best practices from important men in his life—his grandfather; Shahji; and the corn fathers, Mangelsdorf, Brown, and Wallace—Suri led by example in business, working harder than anyone, expecting those in whom he put his trust to be as creative and dedicated to the operation as he was. He had a knack for putting key people in the right positions, creating teams who conducted business with initiative, and striking remarkable deals.

According to Suri's business colleagues and associates, his management style on the surface appeared informal and low-key. He managed people and situations in a disarmingly casual way, but when sharp decision-making was called for, he was acutely prepared and way ahead of everyone.

A tropical corn breeder was hired to maintain and enlarge the capacity Suri had already established at the research station in Jamaica. For the northern European climate, a new research station at Selommes, France, was established in 1972. Hybrids bred at both stations were well adapted to different regions. The Selommes station developed early maturity, temperate hybrids able to turn out good yields in spite of cold, wet springs and relatively cool summers. The Jamaica station, which later moved to Nicaragua, produced tropical and subtropical hybrids with disease resistance and insect tolerance. These were critical for crops grown within thirty degrees of the equator.

Doing seed business in many Latin American countries during the 1970s was problematic because of the government monopolies; regulations often hindered the availability of hybrids to farmers. Skid and Brown approved establishing a joint venture in Nicaragua in 1971 and another one in Brazil in 1972.

The Nicaragua venture was headed by a Canadian named Raymond Gross, who lived with his family in the town of Chinandega. Cotton was an important crop in this area. Suri was shocked by the inhumane disregard for agricultural workers near Chinandega when he saw planes spraying cotton fields, and spraying the workers in the fields as well, with the lethal insecticide Parathion, a highly toxic chemical since banned in many countries. He knew that Parathion "kills everything."

Ray and Suri would eventually do business together all over Central America. Whenever Ray, who was white and fluent in Spanish, and Suri, who spoke only a little Spanish, traveled together, people spoke in English to Ray and Spanish to Suri, assuming that Suri was the Latin. Suri and Ray both found that amusing.

The Brazilian joint venture was initially set up in Porto Alegre, in the state of Rio Grande do Sul, near the country's southern tip. Suri traveled all over the region, scouting out suitable areas for hybrid testing and seed production. The site he chose was on the outskirts of Santa

Cruz do Sul, a German ethnic town about 150 kilometers from Porto Alegre with large tobacco farms, progressive farmers, and a good road connecting the two towns. They built a seed plant and relocated the company headquarters there. A seed production specialist from Iowa took charge of the operations.

To support the joint venture with new and better products, a research station was established in Londrina in the state of Paraná. Suri was thrilled to be able to work again with his old friend and roommate from Harvard, Wertit Soegeng Reksodihardjo. Soegeng!

The economic conditions in Indonesia were so bad that Soegeng couldn't make ends meet working at the Bogor Botanical Garden after leaving Harvard. Suri had hired him first for the Nicaragua station, and now he was transferred from Nicaragua to Londrina to head the new research station. A breeder from the Dominican Republic took his place at the Nicaragua station.

Soegeng's expertise and the research station in Londrina were a perfect combination of needs and fit, even culturally. The region had a predominant concentration of Japanese residents, and Soegeng and his wife, a woman of Chinese descent, each physically resembled the people in their new community. That went a long way in helping to assimilate Pioneer's work there. Another research station at Itumbiara, Goiás, and a second seed plant in Santa Rosa, Rio Grande do Sul, were soon required to keep up with the high volume of sales.

Suri's strategy—introducing high-technology, high-margin products on a cash-and-carry basis to elite progressive farmers working on highly productive land—was the first of its kind in the seed industry in Brazil and soon propelled Pioneer to high profitability. Suri's department was elevated to the status of Overseas Division in 1972, with Suri named as president.

Bill Brown and Suri traveled to the small village of Kapelle in the Netherlands that year to meet with people at D. J. van der Have, a Dutch company best known for sugar beets but also with interest in corn, primarily for silage. Brown and the aristocratic van der Have family enjoyed a longstanding friendship, and the two companies had an agreement in place with no termination date. The terms of the contract gave D. J. van der Have the territorial rights for Pioneer brand seed sales to all of Western Europe. The contract was renegotiated by Suri and Skid to terminate in 1977, when each company would go its own way in the Western European market. To be able to develop that market directly after 1977 was a huge breakthrough for Pioneer.

Suri and Brown drove with van der Have's key people to France for meetings with Pioneer's producer-distributor specializing in corn, SICA France Maïs, a *société d'intérêt collectif agricole*. Suri's counterpart there was Pierre Sarazin, a man who had been instrumental in uniting agricultural cooperatives from all over the country. They met in an historic château in the Loire Valley and observed field trials together the next day. From Paris, Suri and Brown flew to Rome to meet the producer-distributor in Italy, and from there north to Cremona and Parma in the Po Valley to inspect the corn-growing regions that offered potential for developing Pioneer business there.

Pioneer had developed many outstanding hybrids in North America. Pioneer's star performer in the Midwest, hybrid 3780, turned out to be well-adapted in southern France, Hungary, and Romania. Another hybrid from the southeastern US performed extremely well in Italy, Spain, and Portugal.

Suri enjoyed responding to exciting business opportunities, opening new markets, and working with a variety of partners, but he did not enjoy the seemingly endless dinners and obligatory drinking that were involved. Though meals in France and Italy were absolutely delectable, the toasts and conversation lasted for hours and often ended fairly late. He had to get up early the next morning for a long working day, often followed by yet another feast!

Suri traveled to Eastern Bloc countries with a colleague named Miro Jiranek, who spoke several Slavic languages, and he liked to drink! All the habits of Suri's cultural background were at odds with those of Eastern Europe and the Soviet Union where so much drinking and toasting went on, even during the day, before real business was done. But Miro could hold his own drinking with the natives, which seemed to be a requirement for business. Suri and Miro made a good team: Miro handled public relations, and Suri conducted the business.

Brown and Suri traveled to Hungary in 1974 to finalize an agreement Pioneer had been negotiating for the previous couple of years. The same year, they negotiated a contract in Romania, traveled to most of the corn-growing regions, and had meetings with Minister of Agriculture Angelo Miculescu, who was very supportive of the enterprise. On a trip to Murfatlar, near the Black Sea, to celebrate the signing of the contract, Miculescu hosted a delicious dinner in the cellar of the Murfatlar winery famous for its pinot noirs. Suri loved to travel with Brown, and it was always a learning experience. Like with Skid, the collegiality between Brown and Suri was pleasant and supportive. Brown trusted Suri's judgment in negotiations.

As Suri and his team opened up several other Eastern Bloc countries to Pioneer between 1974 and 1981, including Poland, the USSR, Bulgaria, Yugoslavia, and Slovakia, they were received by prominent figures everywhere they went. In Bulgaria, Suri was welcomed by the head of the Communist Party. In other countries, it was a top government official, often from the agricultural side. A common feature of such travels in Eastern Europe was the presence of a Communist Party member keeping a close eye on the proceedings. In dealing with such countries, Suri stayed focused on business and away from politics.

Safety was not an issue on most of the trips Suri made working for Pioneer, but occasionally things were potentially threatening. Suri was in Managua, Nicaragua, with Ray Gross in December 1974 when they heard that a guerilla group affiliated with the Sandinistas had shot and killed the minister of agriculture and seized hostages at a party in his home. Among the hostages were several government officials and relatives of President Somoza. More violence was expected in Managua.

Suri and Ray left immediately for Chinandega, where Ray picked up his family. Together they drove all night on back roads to Honduras. Along the way, they were stopped repeatedly by rebels with guns, just as Suri and Edda had experienced in the Dominican Republic years before. In this case as well, they were allowed to continue on unharmed.

Suri left for the US from Tegucigalpa, the capital of Honduras, two days later. Ray and his family relocated to Costa Rica where he continued to do Pioneer's business in Central America.

Differing cultural customs in business were sometimes delightful as well as mystifying. Suri went back to Warsaw in January 1975 with Miro to negotiate a contract. After reaching their agreement, the company chief invited them to accompany him, saying, "I must show you the wild buffaloes." Miro and Suri had no idea what that might mean.

They set off in a car and arrived late in the evening at a beautiful forest reserve, with snow all around and a luxurious log cabin. As usual, a Communist Party member was one of the five guests and two attendants.

Everyone got into sleighs, traveled deep into a primeval forest, and stopped in a clearing where the attendants built a big bonfire. The high snow had been moved out of the way and piled all around the fire. Polish sausages and other delicacies were served, accompanied by endless rounds of vodka to wash down the scrumptious meal. On this occasion, Suri drank with his associates, though he swallowed many fewer shots than his companions!

The pristine forest was one of the most beautiful places Suri had ever seen: a climax community of ancient trees and associated species of plants. Suddenly, it seemed to Suri, everyone was fully intoxicated. The Poles jumped up and danced joyously around the fire.

When the group exhausted itself, they went on a long sleigh ride, then back to the lodge where the party continued until just before dawn. After a sumptuous breakfast the next morning, Suri and Miro were taken to see a protected herd of buffaloes (European bison). Suri was exhilarated by watching the splendid creatures move with beautiful, carefree energy in that patch of still-wild nature.

That same year, Suri initiated a joint venture on the other side of the world, in the town of Kingaroy in Queensland, Australia. What began on a shoestring budget, in one corner of a rundown warehouse, developed into another highly successful business. This one had the highest gross margin of all the Pioneer operations in the world.

Of the Communist Bloc countries in which Suri did business, the Soviet Union may have been the most difficult. Working with the Soviets would not have been possible if not for the efforts of Bill Brown and his longtime colleague, Roswell Garst, the cofounder of Garst and Thomas Seed Company. These two seed industry stalwarts were invited to the Soviet Union as guests of the Ministry of Agriculture. Garst was one of Henry Wallace's oldest business associates, and he and business partner Charles Thomas were the first distributors of Pioneer seed. His company had been a crucial part of Pioneer's development.

Brown, who became CEO of Pioneer in 1976, and Garst had been able to reach an accord in principle with the Soviet minister of agriculture for the testing of Pioneer hybrids in the USSR. Suri followed up to negotiate the deal. On his visit to Moscow in 1977, a five-year technical collaboration agreement was signed. The two parties agreed that Soviet technical people would visit Pioneer, and Pioneer researchers would visit the Soviet Union as guests of the government.

Around that same time, FBI agents showed up at Pioneer's office to talk to Bill Brown. They also interviewed Suri when he returned from the Soviet Union. Fortunately, the FBI concluded correctly that the motives in Pioneer's relationship with the Soviets were only related to business and had nothing to do with politics.

Whenever Suri had visitors from the Soviet Union or the Eastern Bloc, they wanted to visit Garst's farm in Coon Rapids, which was northwest of Des Moines. Garst gave personal tours of his farm and showed his work on soil conservation and cattle feeding, and his seed plant, which was the largest in the world.[37]

Suri's host in the Soviet Union was a scientist named Domashnev. Among his many admirable qualities, he was a devoted family man. Soviets traveled with very little money when they visited Pioneer in the US, but Domashnev always wanted to take a gift home for his family. He decided on Sears catalogs, which were free, one for his wife and one for a friend.

Suri continued to travel extensively in the Soviet Union, especially in the agricultural regions of Ukraine and western Russia. He and his team explored the country from Rostov to Odessa to Kiev, wherever corn was being grown. These trips brought memorable experiences as well as mutually beneficial agreements.

Eastern Europe was the most profitable and fastest growing operation for Pioneer before Western Europe kicked into high gear.

---

37. As a citizen diplomat during the Cold War in the 1950s, Roswell Garst attempted to promote world peace and American farm interests by reaching out to Eastern Europe with his "Peace Through Corn" approach to exporting hybrid corn seed, resulting in a friendship between Garst and Soviet leader Nikita Khrushchev. International headlines followed when Khrushchev and his wife and children visited Garst's Iowa farm in the fall of 1959. http://www.preservationiowa.org/programs/awardsItem.php?id=140&year=2010.

Everything clicked, and there was tremendous growth. Through 1976–1977 profits grew a previously unheard of 155 percent over the previous year. Suri was always on the move. He hired more staff, but he was still hardly ever home for long. He said, "I was traveling all the time to one place or another, setting up these companies. We didn't care what day it was, whether it was a weekend or a weekday, we were out there working. I was everywhere, because we were expanding so rapidly."

Throughout these years, while Suri was spreading hybrid seeds throughout the world, Edda not only kept the home fires burning and the children safe and healthy, she served as hostess extraordinaire when Suri brought home his international business associates for drinks, hors d'oeuvres, or meals. She took care of the cooking and preparations herself. The kids were often asleep by the time guests arrived for a splendid dinner and lively conversation. But on some occasions, the kids were part of the visit and Edda noticed that the foreign visitors enjoyed spending time with a family.

Visitors from Russia always came with lots of delicious salamis and bottles of vodka in their luggage. At the house, Edda supplied the rye bread and pickles and everyone feasted as they talked and laughed together and toasts were raised "To your grandmother!" and "To your mother!" and so on, one after the other. Edda enjoyed the instant friendships made with these highly emotional Russians who always greeted them coming and going with kisses on both cheeks and warm bear hugs.

That Suri and Edda each had accents from the countries where they were born turned out to be a distinct and unexpected advantage in Suri's international business relationships. He and Edda were told more than once by their guests, from countries all around the world, that people could more easily enjoy and understand interactions with them. Americans usually spoke so fast that foreign guests caught only about half of what was said in any verbal exchange. Those not fluent in English had to first take a second to translate what they were hearing, then guess what they missed before continuing the conversation. Partnerships, friendships, and camaraderie were enhanced by the fact that, to international visitors, these particular Americans were "just like us."

Suri set a strong example at Pioneer, just as Edda did in their home, by treating people with tremendous courtesy and respect. International visitors felt that; this consistent behavior nurtured lasting friendships. As it was in the Punjab, every visitor was an honored guest. As a result, strong bonds were formed, and people were often willing to bend over backward to help Suri when needed. This was especially important in Soviet Bloc countries. Suri said, "You must have a person in every country who becomes your champion. We had lots of them."

As soon as the agreement between Pioneer and D. J. van der Have came to an end, Suri took immediate steps to establish wholly owned subsidiaries in Germany, Austria, Italy, and Spain, plus a joint venture in France. France Maïs marketed the maize hybrids that had been bred by Pioneer and produced by the cooperative societies. The relationship with France Maïs reflected Suri's first rule for starting up any successful new operation: in every country, the important thing is to hire a good person. Pierre Sarazin certainly fit this requirement. Suri moved quickly to put the right people in place in Germany, Austria, Italy, and Spain. Except in Spain, a native general manager headed each entity. These managers shared key common traits; they were bilingual entrepreneurs with a clear vision of how to develop the business.

Suri's overseas research strategy was pragmatic and low-cost. To support Pioneer's markets in Northern Europe, he used Pioneer's proprietary temperate germplasm and combined it with early European flint (hard texture) corn, already in the public domain, resulting in outstanding hybrids with early maturity. Business blossomed and matured, especially once the highly adaptive hybrid called Dea hit the European market in 1978. Dea, a cross between an inbred developed in Indiana and a public inbred in France, was a huge success, miles above the competition. Pioneer was making plenty of money in France. Dea captured a big market share in France, Austria, and other European markets, resulting in an explosion of profits.[38]

A father-son team, Hideo Tokoro and his son, Gensuke (pronounced Gen-skay) Tokoro, were instrumental in helping Suri open up the Asian market. Hideo was a prominent agribusinessman whose

---

38. With the emergence of Dea, profits jumped by 163 percent in 1980/1981 and by 105 percent in 1984/1985.

Ghen Corporation had an ongoing partnership with Pioneer's a poultry division, Hy-Line International. Gensuke had joined Pioneer in the early 1970s, and he now became Suri's point man for Asia. In addition to Japan, Pioneer also established operations in India, the Philippines, and Thailand, and opened markets in other Southeast Asian countries where hybrid corn was something of a novelty.

Gensuke and Suri spent two weeks in China in the late 1970s. As part of a Japanese business delegation in Beijing, they were treated like royalty, with superb meals and many toasts in this case with red wine, beer, and *Maotai* (a liquor distilled from sorghum). Suri presented seminars to Chinese seed experts on technical aspects of corn breeding and the structure of the seed industry in general. The Chinese were hungry for technical information, using these occasions to absorb as much as possible in exchange for lovely dinners.

The late seventies to early eighties brought a profitable joint venture in Egypt and a wholly owned subsidiary in Turkey. Pioneer's investment in Eastern Europe was paying off, especially in Hungary and Romania. Suri attributed the success to high-quality products, good relations with the various government entities, and a novel series of barter arrangements that were the first of their kind in the seed industry. The uniqueness of doing business in Eastern Europe necessitated such creative marketing strategies. Under these agreements, Pioneer's Overseas Division supplied the agricultural institutes in several of these countries with parent seed, from which they then produced commercial seed to meet their domestic needs. Instead of compensating Pioneer in cash, they paid in-kind with hybrid seed, which Pioneer parlayed into cash in the markets of Western Europe. This strategy resulted in dramatic increases in profits.

Other highly profitable projects included converting the joint venture in Brazil to a wholly owned subsidiary of Pioneer, and eventually establishing wholly owned subsidiaries in Mexico, Argentina, and Chile, and a joint venture in Colombia.

Suri's approach to overseas research was to avoid overlapping with Pioneer's domestic research program. All overseas research stations were located in latitudes north of 45° or between the equator and 30° north or south. For those areas of the world where diseases were a great threat, they focused on developing disease-tolerant hybrids. This included breeding for banded leaf sheath disease in Japan, downy mildew in the Philippines, late wilt in Egypt, *Curvularia* in Mexico, maize streak virus in Cote d'Ivoire, *achaparamiento* virus in Central America, and maize dwarf mosaic in Australia.

The big drawback amid all the success for Suri was that, with so much of his attention focused on his work, he had so little time for his family. He was as absent from his family as his own father had been with his.

Suri said, "That's what I didn't do right. I regret not being around more."

Fortunately, Edda filled the void with finesse, just as Shila had for Shahji in the Punjab. The Sehgal kids had full and busy childhoods. All four of Suri and Edda's children were smart; none of them were particularly challenged by the school system. The open classrooms and little homework typical for their schools in those days provided inadequate inspiration. But none of them were ever willing to be put in special classes for gifted and talented kids. Like Suri and Edda, they wanted to be just like everyone else. With parents who had accents, they wanted to be "super" American.

When he wasn't playing with friends, Kenny was still buried in a book much of the time. He liked fantasy novels and science fiction. He was sentimental and more serious than his siblings, according to Edda, though not very disciplined. He waited until the last minute to do his schoolwork, and then only completed the bare minimum. Kenny was good at building ramps and forts and organizing activities with other kids—BMX bikes and skateboarding and stickball games in the summer, sledding and snowball fights and football in the winter. He loved music and played trombone in the jazz club at school. Acutely aware of injustices around him, he befriended a kid with a huge temper and "some kind of bipolar problem," who sometimes "freaked out." The boy played snare drum in the marching band, which calmed him down. Other kids in school made fun of him, and Kenny was appalled that even the teachers picked on this boy, too, trying to further rile him up. Kenny took him aside occasionally to deflect the negative attention. But every time Kenny talked to him, he had to put up with kids in the jazz club "ragging" him about it. They didn't understand why he was nice to this "complete freak."

Bernd, who by fourth grade was going by the name Ben, was the quiet one, reliable, well-spoken, and thoughtful. Edda thought he was more German than she. His brothers sometimes teased him about his self-discipline and thorough attention to detail. An "A" student who always did his homework, Ben was able to do well in the open classroom system that Kenny disliked. Ben was immersed in sports, especially baseball, which he played every chance he got. He loved music, listening to AM radio and later the local rock station, KGGO. He played drums for a couple of years in grade school, but balked at the early-morning practice schedule.

Oliver was full of ideas, an instigator, outgoing and popular with the neighborhood kids who often rang the doorbell, looking for him. He did well in school and played well with Vicki. His asthma never stopped him; he played hard until he literally collapsed at times. All three boys played baseball in little leagues, where Edda volunteered at the concession stands to watch their games. Around the neighborhood, the Sehgal boys had little trouble drumming up a pickup game of stickball or tackle football.[39]

The boys were not too interested in playing with their little sister. They thought Vicki was cute, but she was too young, and a girl. They played rougher games, and she sometimes got in the way and interrupted their fun. They nicknamed her The Pest.

When Edda forced the boys to play with Vicki, they devised "jobs" for her to do, such as protecting the sewer drain so their balls didn't go down it. She watched for cars when the guys raced down the hill on their bikes or skateboards, and she made sure nobody was watching when they were doing something they weren't supposed to be doing. She had a bucket of water on standby when the boys set fire to their toy rockets.

Vicki was thrilled to be included, and she enjoyed the company of her big brothers. She was the grounded, responsible one from an early age; her observant nature and natural resilience no doubt helped develop

---

39. In the absence of other kids, the Sehgal boys could still play a three-person football game they were sure they invented called "pass-catch-defend," in which a player gets a point for each touchdown throw, touchdown catch, or interception.

the skills for her potential future as a primary school teacher—a career she decided from her earliest childhood that she would pursue—if she didn't become a pediatrician.

Vicki and Edda were always close. Edda said, "Vicki thinks like me, with my interests and tastes. She's special. Suri was always soft with her. If he even raised his voice, she started to cry. The boys were much tougher. Vicki got along well with them, and they adored her."

There were days when Edda felt the pressure when Suri wasn't around; she sometimes became stressed out by the weight of so many responsibilities while also being surrounded by four rambunctious children. She became impatient with the boys more frequently, because they tended to get on her nerves and get into trouble. Kenny didn't participate much in destruction or vandalism, but Ben and Oliver carried out lots of typical boy pranks. Ben was more careful, and only Oliver got caught, even when they were doing the same thing. The police came knocking on the door after the boys wrote in wet cement for the third time at a nearby construction site; writing "Oliver" in wet cement was a dead giveaway when there was only one Oliver in Urbandale.

The kids joked that Edda could never say, "Wait till your father gets home!" if they misbehaved, because it could be late at night or even two weeks later before Suri got home. So Edda ran a tight ship. She encouraged the kids to play outside most of the time during the day. During the school year, they came home after school, had a snack, and sometimes watched a little television. They did their homework, if they had any. The rule was that they had to be inside for dinner at 6:30. After dinner, television was allowed from 7:00 to 8:00 p.m., family shows only. They were in their rooms by 8:00 each evening, and lights out at 10:00 p.m.

Edda wanted peace and quiet in the evenings. If Suri was home to eat with the kids, which happened occasionally, he enjoyed the company of his family and never talked about his work. He read while the kids watched TV after dinner. When Suri came home late after work, he and Edda ate dinner together later in the evening, and Edda apprised him of the things that had gone on that day with the household and the kids.

The Sehgal home in Iowa was not at all extravagant. The Midwestern disdain for showing off matched Edda's deep-rooted German background. So the kids had no idea that Suri was so successful in his work. There were hints, of course. No one else in Urbandale had important visitors showing up from foreign countries, let alone Communist ones.

Suri traveled a lot more than other dads and made plenty of international phone calls. He came home with interesting things from his trips, things other kids didn't have, such as Russian dolls and African woodcarvings. And though their house was modest, it had particularly nice carpets.

Suri brought stamps, coins, bottle caps, and beer cans from countries their friends hadn't even heard of. Their pals had Budweiser cans, but the Sehgal boys had cool ones: Carlsberg, Tuborg, Henninger, Foster's, Tres Equis, Brahma, and others. Suri sent the kids postcards slapped with as many different stamps on the envelope as he could fit. The kids soaked them in water to remove them. The kids had hand-held Nintendo games and a Sony Walkman, courtesy of Suri's Japanese business associates. None of their friends had those unusual things.

Wonderful packages also came regularly from Edda's mother, full of things the kids never otherwise saw in Iowa. Oma Margarete sent lots of German chocolate, cookie treats, and the best muesli, plus Legos and other European toys.

The boys were approaching their teen years before they started figuring out that their family actually came from very different backgrounds—and that meant they were a bit foreign as well. Their parents each represented something exotic. While they were young, the kids did not have the sense that their parents were that different than anybody else's, even though their dad was slightly darker in complexion, and both parents spoke with accents.

As much as Edda tried to be as American as possible, she sprinkled everyday life with her German roots in little ways, such as serving pancakes dipped in sour cream instead of maple syrup. Nobody the kids knew ate pancakes that way.

Vicki was teased occasionally at school because her parents had such uncommon names. When Edda slipped some German chocolate into her lunch, Vicki was likely to trade it for some Little Debbie treat. Most kids ate sandwiches on white bread. Though Vicki liked the brown pumpernickel bread Edda used, Vicki didn't ask for it often because other kids commented on it.

Once a year, Suri took the two youngest, Oliver and Vicki, to spend a day with him at work at Pioneer. They played in his big corner office, hounded his secretary for inkpads and paper, and played with the telex. All four kids went to his office occasionally on weekends.

Though the kids may not have had a clear sense of Suri's position at Pioneer, they could see that Dr. Brown was an important person. The corporate culture of Pioneer led by Bill and Alice Brown at the time was a close-knit family network. Everyone at Pioneer spoke often and warmly of Dr. Brown: the tall, gentle man the kids knew as a grandfather figure who smoked a pipe and spoke slowly, always taking time to chat with them about whatever they found interesting.

All the Sehgal kids loved going to the Browns' beautiful home, which was surrounded by trees, plants, and a wild garden on a large wooded lot. Their house was full of antiques, including Alice Brown's loom, where she weaved items she sold in a little store in the suburb of Beaverdale. She always gave the kids books and other presents for Christmas and on their birthdays. All the kids loved playing games with the Browns' granddaughter, Lila.

Oliver said, "Everybody at Dad's work talked about Dr. Brown, and we would say, 'Well, we know him. We've been to his house!'"

CHAPTER 14

Edda's Mastery

Edda's talent at providing a welcoming home to others was far more fully realized when family members from India started to arrive in Des Moines. Still deeply concerned because opportunities for young people in India, even educated young people, were nonexistent at the time, Suri had offered to help his relatives migrate to the US for a chance at a better future. He had arranged for two of his nephews, Shakuntla's son Chander and Kedar's son Ranjiv, to go to school in Japan. Chander did well attending Sophia University in Tokyo, but Ranjiv returned to India.

Kedar was increasingly concerned about Ranjiv, now twenty, and another of his five sons, sixteen-year-old Rajat. Neither boy seemed to understand the importance of education in making something of themselves. Kedar had been slow himself in realizing the value of a good education, and only after much prodding from his father, did he get his undergraduate and graduate degrees in his thirties. But Kedar was not as patient with his sons as Shahji had been with him. Suri suggested that Kedar bring the two boys to live with Suri and Edda and their four kids in Iowa. He would do his best to mentor his nephews.

Kedar, Ranjiv, and Rajat flew into Chicago in May 1977, and Suri and Edda picked them up and brought them home. Kedar stayed on for a few weeks before returning to India.

Edda took charge immediately, acclimating the boys to their new home while Suri was at work. Little Vicki, barely four years old, showed her Uncle Kedar how to do and say simple things. He adored her. The family traveled together to Omaha to file for Kedar's reentry papers, and they took drives to various family places of interest on weekends when everyone could be together. On a weekend outing to watch the annual US National Hot Air Balloon Championships in Indianola, Iowa, the little kids were having trouble pronouncing Rajat's name, so Edda said, "We need to pick a new name for you!"

Nine-year-old Ben[40] suggested calling him Jay, and Rajat liked the sound of that.

Jay had arrived as a fairly pouty teenager. He was so shy he couldn't speak to adults or young women. He had no interest in school. He had done schoolwork when necessary, but he dreamed only of becoming a cricket star. He had not been close to his own family. His father's disapproval had occasionally been expressed in the form of physical punishment and, as a result, Jay had a keen sensitivity to self-protection. In India, he always knew he was safe if his grandfather, Shahji, was around. Shahji would not allow anyone to raise a hand to their children. Jay could see right away that he had landed in a safe place with Suri and Edda. They were a loving team of surrogate parents.

Jay was most comfortable immediately with Edda, who understood how he felt as a teenager in a completely foreign environment. She wanted him to feel at home. Jay remembered Suri's visits to India, and his visiting Ludhiana to give a talk at the university. Uncle was such a great figure in Jay's mind that he was a little scared of him.

Steady and self-disciplined, Jay didn't take easily to change. He was the opposite of his brother in personality. Ranjiv was confident, ready for action, and anxious to be out on his own. They each had a room in the finished basement that had been Kenny's and Ben's rooms. Kenny moved upstairs with Oliver, and Jay shared Ben's bunk bed. Ben and he were closest in temperament, and they both loved sports.

Suri enrolled Jay and Ranjiv in English classes at Des Moines Area Community College and helped both brothers find part-time jobs. Ranjiv worked at a gas station, and Jay worked at McDonald's.

---

40. Ben was familiar with the dilemma caused by having an unusual name. Before he took on his own nickname, Ben, it seemd that no one pronounced his name correctly. His friends called him "Burned" or "Burnt."

Though Jay could read English, he refused to speak it. Suri told him firmly, speaking to Jay only in English, insisting, "No more Hindi or Punjabi. You have to speak English from now on!"

Jay rebelled Indian style: he became upset and refused to eat. In India, when a child didn't eat, parents became upset. But no one in Suri and Edda's home cared if Jay chose not to eat; no one catered to his moodiness. He quickly abandoned that method of rebellion.

Edda could see that Jay needed more gentle attention, and she talked to him every day, advising and mentoring him. Jay credits Edda and Suri with his ability to get through the initial turmoil of adapting to life in America, and with his eventual maturation. He said, "I am sure I gave them nightmares."

But Edda never thought Jay's presence in the family was any problem. She said, "He had never eaten the foods we ate, but he got used to them. He was sometimes a little sensitive. If something went wrong, he might get a stomach ache. If the boys teased him, he had a bit of a temper. He'd leave the dinner table and slam the door, but he got over it. Such new surroundings were a big adjustment. He was a bit rebellious, and his brother was no help."

Ranjiv was not an ideal role model. He moved out without saying good-bye by the time Jay started high school in the fall. Jay came home late from his evening job that summer to find a note left on the table in the basement. Everyone else was in bed. For Ranjiv to leave like that upset Jay. Such self-determined autonomy was an alien concept to him. Distressed and anxious, he woke up Suri. Suri made some phone calls and located Ranjiv.

Suri told Jay not to worry; Ranjiv was happy to be on his own now. Kenny moved back downstairs to his old room, and Edda continued to run an efficient home operation.

Jay knew that Ranjiv was basically a good-hearted guy whose preference for instant gratification in all things got him into frequent financial trouble. Jay gave his brother his McDonald's earnings from time to time when he needed help, but Uncle Suri was the one who bailed him out most often.

Jay got up early each morning to iron his shirt and pants. That gave him one-on-one time with Uncle, because Suri was also up early, drinking tea and reading the paper. Talking with Jay before going to work,

Suri commented on whatever news he'd heard from Edda the night before and offered guidance and encouragement. Jay benefited greatly from the pep talks and personal attention. Jay said, "Though Uncle was gone a lot, when he was home, he was there one hundred percent."

Uncle, Auntie, and the four young cousins embraced Jay and accepted him completely. He did well in school, especially in math, and was given some extra help with reading and pronunciation of his new language. He kept the job at McDonald's all through high school.

Though Jay never had an opportunity to play cricket in the US, he enjoyed playing soccer during his senior year of high school when the sport started to become more popular in the US. He played some baseball and joined the kids' pickup football games, impressing them with his strong, cricket-honed arm. He took drivers education classes, and Suri taught him the behind-the-wheel portion.

Jay came home one day during the summer of 1978 to find a note on the kitchen table. Edda had received a call from India. Suri's father had died of heart failure. Jay read the note with alarm, realizing his grandfather was gone. Edda had immediately called Bill Brown because Suri was in Romania, and she didn't know how to reach him directly.

Suri received an urgent message from Bill Brown to call him. Suri learned the sad news of Shahji's death from his mentor, his other father figure. Suri arranged for a flight the next day to India.

The custom in India to have the body cremated within twenty-four hours meant that Suri would not get there in time for the actual funeral procession, but there would be a full seven days of memorials, eulogies, *bhajan* (songs), pujas, and other devotional practices to honor Shahji.

Jay's brother, Raman Sehgal, was twenty years old when Shahji died. He had spent his childhood with his grandparents in Amritsar. Sometimes, on school holidays, Shahji had taken Raman along on his out-of-town travel to attend meetings or visit government offices in the state ministry at Chandigarh. When Shahji died, Raman was in college at Ludhiana.

Raman was one of almost 200 people in the procession that day as Shahji's body, dressed in white and covered with flowers, lay on a wooden plank and was carried on the shoulders of family and friends to the cremation grounds. Raman was astonished that he hardly had a chance to carry his beloved grandfather very far because so many people wanted to help carry him.

Upon arrival at the grounds, at least 200 more people were waiting to pay their respects to Shahji. The carriers took twenty minutes just to weave through the crowd to where the Hindu priests would perform the final prayers and blessings.

During the long walk in the heat, Raman was startled to see his grandfather's nose start to bleed, something that happened often in summer throughout Shahji's life. For a brief moment, Raman hoped his dear grandfather was not dead after all, but he was.

The morning that Suri landed in Delhi, Sanjogta's husband, S.K. Kapoor, met his plane and took Suri to their home. Sanjogta was already in Amritsar. S.K.'s father also wanted to attend the puja ceremony that was being held at the house until the seventh-day *kirtan* (praise eulogy) ceremony. S.K., his father, and Suri flew together to Amritsar and participated in a few days of puja.

The kirtan on the seventh day included religious music and speeches. Expectations were for mourners to gather in the courtyard of the Sehgal home, where Santosh and her husband had been living with Shahji. Because such a large crowd had attended the actual funeral, the Sehgal family did not expect many to come to this ceremony. But so many people were arriving, it was clear that the courtyard would not be large enough to accommodate everyone. The entire street ended up being blocked off for this special occasion. This crowd was far larger than the gathering at the funeral because many people had to travel long distances to get there, just as Suri had. Shahji had helped so many people in his life and was close friends with so many. They wanted to show their respect.

Shahji's lifelong friend and fellow activist, Pandit Amar Nath Sharma, orchestrated the eulogy. Many people had come to say a few words about Shahji—local politicians and community leaders, a Sikh member of parliament, ministers from the Punjab government, people from schools Shahji was instrumental in forming—associates and friends from all walks of life. One person after another expressed their heartfelt love, respect, and gratitude for the life of Shahji Sehgal.

Raman remembers seeing his Uncle Suri standing near the dais, filled with emotion and tears in his eyes. The memorial was a fitting tribute for a man who meant so much to so many. Shahji was a heroic figure to many in his country, his community, and his family.

Raman described his grandfather: "He was a great person with huge patience and very polite in his speech. He went out of his way, without selfish motives, to help others. I have always turned to his values in times of difficulty and think how he would have handled a situation. He lives in my heart."

When Jay finished high school in Urbandale in 1979, he told Suri that he wanted to work for a year, study politics, and go back to India. He was bothered by the typical view most Americans seemed to have that India was nothing but a poor country. He wanted to change that perception and make a difference. He had an idealistic view about politics being the way to go.

Suri told him, "You can do that, but you will starve."

Suri was not enamored with politics, considering it "a dirty game with too much corruption." He suggested that Jay instead go to college right away and learn a skilled profession.

Jay decided to take the year off anyway, working at McDonald's full time and at a second job at night as a security guard at an amusement park called Adventureland. He moved into an apartment with his cousin Rick (Savitri and Brij Anand's son), who was working toward a pharmacy degree at Drake University. Suri continued to emphasize the need for a college education and finding work that would sustain a lifetime, but he did not interfere with Jay's choice to take the year off.

Each summer, Pioneer celebrated the Fourth of July and Pioneer Hi-Bred Day by treating their employees to a big picnic at Adventureland. The Sehgal kids loved going there even more with Jay working there. They rode roller coasters and played games. If Pioneer visitors were in town, they all went to these events, or to the huge Iowa State Fair, which was held over eleven days each August and attracted people from all over the world.

Sehgal family vacations during the summer were a tradition throughout the seventies and into the eighties, usually with a huge packed station wagon and a memorable destination. Experiencing the American West for the first time, the kids were especially impressed with the natural panoramic locations, such as Pikes Peak and Royal Gorge, which were so different than the rolling plains and rivers back home in Iowa. Bill Brown offered the use of his duplex nested in a secluded area of woods that backed onto a stream in Winter Park, Colorado. The kids liked to play in the stream, hopping from rock to rock.

On a family road trip to visit friends in West Virginia in the summer of 1979, the Sehgal kids were introduced to the music of the rock band Kiss. Oliver, not yet in third grade, was an instant fan of the group's flamboyant style, black-and-white face paint, and over-the-top creature costumes and antics. The boys collected Kiss albums, joined the "Kiss Army" fan club, and became very excited when they learned that the Kiss Dynasty Tour was coming to Veterans Memorial Auditorium in Des Moines that October. Oliver led the effort to convince Suri to let them go to the concert.

Suri could not see the appeal of the group, but he finally gave in and agreed to take them. Edda suggested taking along Jay, now eighteen. The five Sehgals were not the typical Kiss audience members. Oliver had just turned nine, Ben was eleven, and Kenny was almost thirteen. Most of the crowd enjoying the performance appeared to be older teenagers, many of them intoxicated or smoking joints openly during the concert.

Suri, Jay, and the boys, standing near the soundboard, had a good view of the stage. Looking around the stadium, Suri, at age forty-five, became aware of being the oldest person there. He had a sincere wish to be somewhere else—almost anywhere else.

Watching the pyrotechnics, blood spitting, and loud explosions on stage, the boys were having a blast. But Jay's jaw dropped to be standing next to his somewhat austere uncle in these circumstances.

After several minutes, Suri turned to Jay and said, "So this is how you make money in America?"

Looking back years later, all the Sehgal boys would consider that experience hilarious.

Meanwhile, Jay was acquiring some firsthand insight concerning the relationship between education and making a good living. He was shocked to realize during this year off that even working up to eighty hours a week at his two jobs was still barely enough to live on. He looked at his older brother Ranjiv—always spending more than he earned. Jay's eyes were now open, and he said to Suri, "Uncle, I've had enough. Flipping hamburgers is not a way to earn a living. I'm ready to go to college."

Suri said, "Good! Let's start applying to colleges."

Jay started at the University of Iowa the following fall.

His presence in the Sehgal household had offered the kids a more intimate entrance into, and understanding of, the complexities of Indian culture. This was multiplied again and again as Suri and Edda made it possible for more family members to immigrate to the United States to live and/or study in Iowa.

In the spring of 1980 "the aunties" and their families started arriving. When Suri was away, Edda was the primary host and helper to the ever-growing extended family.

Vicki was six years old when the first family from India came to the US; Suri's sister Santosh (called Toshi) and her husband Sudershan Sabharwal (called Wally) came with their nine-year-old daughter, Nancy, and teenage son, Sarat. Vicki loved showing her relatives everything in the house, and teaching them how to do things. She and Nancy became good friends.

Suri and Edda rented an apartment for the family in the nearby town of Clive, provided them with a car for mobility, and took Wally around to find a job. The apartment complex had a swimming pool, and the Sehgal kids spent a lot of time there that summer. Edda took the girls for swimming lessons.

Padma and her son Sanjay came that November, as well as Parsanta, her husband Karanjit Puri, and their three children, and later Padma's husband. By the time Jay finished college, with a double major in computer science and business administration, there were at least twelve Indian relatives all living nearby, each looking to Edda and Suri for guidance.

Relatives came with only what they could carry in their suitcases. Some had language problems, and everything was foreign to them. Even simple things such as doing laundry, shopping, or enrolling kids in school took a lot of effort. Cooking was not a problem, but finding

Indian groceries was challenging. There were no Indian grocery stores in Des Moines. They were not used to washing machines, and no one drove except Wally. Suri had taught Wally how to drive a few years earlier when he'd visited Des Moines. Driving was a necessity in Des Moines.

The boys who came were already teenagers, and most of the girls were in the upper grades of elementary school. There were three boys and three girls, and the Sehgal kids noticed that the girls didn't play with the boys.

Edda welcomed her sisters-in-law and their families, and served as the go-to person for each of them, just as she always had for Jay. She was the chauffeur who shepherded them through the agencies and bureaus, helping them get medical care, dentists, and necessary licenses. Vicki was delighted to help with spoken English and in explaining various details of daily life.

Suri had a definite philosophy, based on his own experience, of how his relatives had to acclimate to American life. He explained, "As a foreigner, you start with many disadvantages. You look like a foreigner, you talk like a foreigner. You must make an effort to become part of the mainstream."

His advice was very simple: stand on your own feet. He felt that when coming from a country like India, it was too easy to accept being dependent. In America, independence is a must.

Suri was pragmatic, firm, and demanding in his approach to his Indian family members. He was committed to getting certain things done without delay. Each person had to speak English, learn to drive and get a driver's license, and go to school or get a job—any job, and eventually become self-supporting.

Suri provided each family with a place to live, a car for mobility, and financial help to get started. All this was dependent on each person's needs—he did not consider his help a handout. He wanted his sisters and their families to be on their own so they could begin to adapt to American ways as quickly as possible without having to worry about basic necessities.

Iowa's governor had welcomed many Vietnamese refugees, and the state was paying a minimum wage to refugees to attend ESL (English as a second language) classes at the local community college. Being paid for six months to polish their English skills was a big help for people settling into a new country. Suri enrolled his sisters in the classes and insisted that they speak in English, just as he had with Jay.

Suri's Indian family members sometimes found the challenges he thrust upon them frustrating and difficult. Money was tight. Finding comparable employment as they had in India was tough for the men, and the women were not used to having paying jobs. Without skills, some took jobs in hospital kitchens, which they found degrading at first. But they soon realized that no job was degrading in America.

The beginning was the hardest. They were all homesick, but they managed with the safety net they had in Suri and Edda. The high school boys got jobs in fast-food restaurants, which contributed to the family finances. They would each go on to college and end up with good jobs. They all went on to earn advanced degrees, some becoming medical doctors. The parents worked long hours and made sacrifices for their children's future, and the children took it seriously; they studied and worked hard, embracing the American way with Edda and Suri as their role models.

From their first meeting in India, the whole family had loved Edda, and she loved them. She noticed now that if they wanted to get to Suri, they often went through her, assuming she had greater influence. She thought Suri was sometimes too blunt with his advice for his sisters. She explained, "Suri has a special place as the guiding force behind his family. But everyone was a little wary of him. Like a strict father, he could be a little rough, outspoken, not very tactful at times. He had so much patience and tolerance with his employees, but with his family he was firm and less patient."

That Padma's younger brother would no longer speak to her in Hindi or Punjabi was a shock. She observed many years later, "That was hard, but looking back, it was the right way."

Padma's son-in-law Ashok agreed. "The first year was tough, but with Uncle's and Auntie's help, things went well. They gave us guidance and full support. We really thought Suri was pretty authoritarian. However, in retrospect, we realized that if he didn't do that, half of the people would not be where they are now. He taught them discipline, taught them to do things. Of course, it was sometimes harsh, but that's the way to get a message across—for people to achieve what they are capable of achieving. Uncle believed, and enforced his belief by example, that language, education, and good work ethics are essential for competence in the American workplace. In short, adapt as quickly as possible, and then follow an educational track to upward mobility."[41]

Welcoming so many family members from India was an adventure for the Sehgal children, who adapted to these colorful newcomers in varying ways. Edda's parents had also visited, and the children returned those visits and were getting to know their German grandparents as well. But India was new—and much more foreign than Germany. Suri and Edda's kids began to realize that their father was not as representative of India as these newcomers were.

The long-term presence of this extended family introduced unusual elements in diet and other cultural practices. The kids learned how to eat Indian food as the families shared dinners. The scent of curry filled the halls of their apartment complex. And these relatives dressed in saris on special occasions. The men often wore traditional outfits at home, observed their religious rites in little shrines set up in their houses, and had puja rituals.

Oliver and Ben cringed when their vegetarian auntie, in an attempt to integrate into American culture, ordered a Quarter Pounder at McDonald's without the meat—just getting the bun, cheese, and toppings. Things like that could be somewhat embarrassing for the kids.

Some cousins were more or less the same ages, and they all got along really well, playing cards, games, and basketball. Vicki especially embraced her Indian relatives, becoming close friends with several of her girl cousins. Some of the younger cousins came along on the family's road trips, as Jay did.

Padma and Parsanta and their families eventually lived side by side in a duplex Suri bought for them in Clive. They had a big backyard with a vegetable garden in a neighborhood with good schools. The Sehgal kids spent lots of time there for dinners and various celebrations.

---

41. Padma arrived in Des Moines in 1980 with son Sanjay. Padma's husband Ram Chatrath came to the US a year and a half later with their daughter Ladi, the oldest of the girl cousins. Ashok followed—his marriage to Ladi had been arranged in India. Ashok graduated from Northwestern University in dentistry. Padma and Ram eventually relocated to the Chicago area after Ladi and Ashok settled there.

In time, all the relatives would become independent. With each other for support, help from Suri and Edda was needed less and less. However, their social contacts remained warm and wonderful. As it was in the Punjab, there were many gatherings and celebrations.

The families maintained their cultural roots and rituals and adopted American traditions as well. There were many get-togethers, birthdays and graduations, and later engagements and marriages—most of them arranged and some not.

CHAPTER 15

Karma

In dozens of countries around the world, Suri confronted all kinds of situations in surprising combinations of the unexpected, an endless variety of people and cultural differences, operating in diverse political and business environments. A mixture of intuition, serendipity, constant preparation, attention to detail, and genuine goodwill offered him extraordinary opportunities. His skill set was honed in an astonishingly varied and wide theater of experience. But as comfortable as Suri was in those international business situations, back at home in the Pioneer executive suite, he was no office politician. That became poignantly apparent in the early 1980s.

In all of the successful business relationships Suri created for Pioneer, in what continued to be an enlarging portfolio of achievements, Suri either led the negotiations or sat in on them. His technical background helped a lot. He found it fun to watch people negotiate. He adapted his style to certain cultural aspects he found in different countries. He noticed that the French seemed to always make straightforward things more complicated. In contrast, the Dutch had a way of keeping interactions simple; they knew exactly what they wanted and usually got it in the end. With the Italians and Latins, everything was emotional. Flexibility and accommodation were needed with the Soviets and Eastern Europeans; and it was important to work top-down, not bottom-up, in those countries, because their flexibility was limited. If Suri had an understanding with the top man, matters went smoothly.

With the Japanese, building trust over a period of time was most important, and he found the Chinese to be the fiercest negotiators.

Suri's negotiating style was to be very clear about what he wanted to achieve, what his positions were, setting the agenda, and deciding who was to do the talking. He told his team, "Never go into a meeting without a very clear idea of what you are trying to accomplish. It's a must." Respect and courtesy were fundamental and well understood by every member of his international team.

Suri kept his options open; he didn't close doors or say that anything was impossible. He made sure he had a range of alternatives. At most, he might admit that someone's proposal might be difficult, or could create a problem—but he never ruled out any idea. By exploring possibilities, he was able to move the give-and-take gently in his direction, always striving for a win-win outcome.

On top of every detail in negotiations and with any planned events, Suri trusted his people to follow through where it counted. Those who didn't measure up were weeded out quickly. The team functioned like a well-oiled machine and a close-knit family. Each team member was called upon to perform to the greatest advantage. Such was the case when Suri invited Peter Dewald to sit in on negotiations with Eastern Bloc countries.

Dewald, born in Czechoslovakia, was fluent in several Slavic languages. Each year, just before planting season, Suri and his team got together with the seed production people in each of the Eastern Bloc countries where Pioneer was producing seed—Romania, Hungary, and Yugoslavia. During the Berlin Wall era, there was always some tension crossing borders, so the meetings were held in Vienna.

Negotiations were orchestrated over several days, with one group per day. Dewald sat in on the discussions about the acreage they needed, what hybrids to produce, or how much to produce of each one. During breaks, Suri and the team got together, and Dewald's understanding of the languages helped the team better understand what the other side talked about.

The international business was doing spectacularly well by the time Bill Brown retired as CEO of Pioneer in 1981. Tom Urban became the new CEO. The name Overseas Division was changed to Pioneer Overseas Corporation (POC), with Suri still serving as president.

To honor Bill Brown's extraordinary contributions to the success of Pioneer, Suri and his team threw a spectacular retirement party for him in Honolulu, Hawaii. The stunning event was integrated with POC's tradition of hosting celebrations at five-year intervals that included all the international partners. Suri felt it was important to gather Pioneer's key people from around the globe to meet each other, learn from each other, and celebrate their successes collectively. Excellence was the trademark of these meticulously planned affairs.

Suri had Gensuke Tokoro coordinate the event. He added several Japanese touches to the party: flowers everywhere, gifts for everyone, and a beautifully decorated hospitality suite, well stocked with snacks and drinks, overlooking the Pacific Ocean. Gensuke was a very detail-oriented person, so he rehearsed until everything was just right. He set up a small office in the hotel for anyone who wanted to send a telegram or telex or make a phone call. The program for the next day was printed in the evening and delivered to each room at night under the door. The organization was flawless.

During the day, there were one or two business sessions; the rest of the time was free for attendees to play tennis, golf, or just relax, enjoy drinks, and take in the gorgeous scenery. Programs were offered for spouses as well. Every evening before dinner, people assembled in the hospitality suite for happy hour. Gensuke had assigned POC people to tend the bar.

It all led to the banquet to honor Bill Brown. The occasion was filled with warmth and good feeling. Bill Brown had been a charismatic and soothing figure at Pioneer, and a guardian of the company values. His departure was a true loss to Suri and to the company.

CEO Tom Urban and other leadership folks from the domestic divisions attended, as did the van der Have people from the Netherlands, and associates from Hungary, Romania, Italy, Brazil, Australia, and many other places around the globe. Forty countries were represented.

Alice Brown spoke about life with Bill, his involvement with the Missouri Botanical Garden, and his extraordinary career in business and science. Si Casady, who had been with the company since the beginning, made a speech on the history of Pioneer.

In his laudatory remarks Si included praise for Suri and the resounding success of the Jamaica project, saying, "Dr. Brown had a dream (tropical corn breeding). And Suri made that dream his own. He went to Jamaica and seemingly dropped from sight with only the meagerest support from Des Moines." Si described how Suri was able to work with the "cast-off resources of less imaginative men." He credited Suri with the energy of a steam engine and with the great success that followed.

But the obvious respect and affection everyone conveyed toward Bill and Alice Brown and Suri that night in Hawaii did not appear to sit well with the new CEO. He maintained a somber facial expression throughout the festivities.

At the end of the evening, Suri asked Urban, "How did you like it?"

Urban scoffed, "It was like the emperor's wedding."

In retrospect, Suri said, "In the enthusiasm of honoring Brown, we neglected to pay enough attention to Urban. I regret that. After all, he was the CEO."

Suri thought that Tom Urban was basically a decent fellow who wanted to further build Pioneer Hi-Bred International, Inc., and that he supported the founders' model of a lean corporate structure with several decentralized business units. But a transformation in another direction was about to occur that was engineered in large part by Urban's chief financial officer. Centralization of Pioneer's operations and the divestment of the company was the long-term plan. The profit centers that had grown Pioneer's success were to be eliminated; a central structure would be imposed.

Urban and his CFO had already cut off relations with the Garst and Thomas Company of Coon Rapids, Iowa. This had resulted in a bitter legal fight between the two companies, whose affiliation had begun decades earlier with a handshake between Henry A. Wallace and Roswell Garst, two of the most highly respected figures in Iowa history and in agriculture. Garst and Thomas lost the legal fight and separated from Pioneer, which generated animosity and rivalry that soured what had been a long and successful relationship.

As a fierce defender of decentralization in the international operations, Suri was now at odds with the corporate office. He was considered part of the "old management" of Brown and Skidmore. Suri was convinced that centralization was not the right model for Pioneer's international business. The Cold War was continuing and numerous trade barriers still existed. He naïvely assumed that, as long as POC continued to do well and kept growing, the new direction was not a true threat to the overseas team.

However, Urban took steps that drove a deep wedge into their relationship. He instituted a formal audit of POC, looking for any signs of mismanagement. A legal team from Washington DC spent more than two weeks going through all POC documentation. They found nothing inappropriate and their report to the board was positive. Urban then brought in a former associate as a consultant to suggest structural and strategy changes in POC operations. These incidents, one after the other, created mistrust and insecurity within Suri's independent team, which had developed and flourished in a protected environment that had always been shielded at the top by Brown and Skidmore. But with those leaders gone, the safety net was gone as well.

Further incompatibilities became obvious when Suri traveled on business accompanied by Urban, as he had done with Skid and Brown on many occasions. Some of the hallmarks of POC's business style were missing in the new CEO.

Most commonly on Suri's business trips, even to places he had been visiting for years, he arrived after a long flight somewhat tired and ready for some rest, usually with jet lag and a several-hour time change to adjust to. However, his hosts had inevitably put together a full schedule of meetings, presentations, leisure goings-on, and activities to showcase their country's tourism features, their work, and their collaborative friendship.

Suri understood his hosts' desire to make a good impression, and he was always gracious and accommodating in expressing his full interest and appreciation for everything he was presented with, regardless of his level of fatigue. His down-to-earth, nonthreatening personal style was a key factor in developing mutually respectful and warm relationships around the world. CEO Urban came across far more officiously, speaking only of his own work, Pioneer, and "how we do things in America."

The first time Urban traveled with Suri to Hungary, their hosts were people Suri had been dealing with for a long time. A great deal of familiarity and obvious warmth was expressed toward Suri, which

Urban did not appear to appreciate. The situation between Urban and Suri felt tense as people addressed all their questions to Suri rather than to the "new guy."

When Urban and Suri traveled to Romania in the mid-eighties, Suri's old associate, Angelo Miculescu, now the Romanian ambassador to China, had already made arrangements for Urban and Suri to meet the Romanian "great leader," Nicolae Ceauşescu, to obtain an endorsement of the project "from the top."

Any meeting with Ceauşescu was top secret, and security was tight. Suri had to wait with Urban in their hotel for a couple of days before they were taken to the palace. Suri understood that it made no difference whether they exchanged any information of substance with the leader or even spoke with him at all—it only mattered that such an encounter occurred.

The fifteen-minute meeting with President Ceauşescu was cordial, and the president was supportive of the collaboration between Pioneer and Romania. Urban seemed comfortable within a clearly defined power hierarchy.

Later in the day, Suri and Urban went to the National Agricultural Research and Development Institute at Fundulea, east of Bucharest, to meet the top guy there, who began, "Now that you have met our president, we can seriously discuss our cooperation."

The Pioneer international meeting in 1986 was held at a resort hotel in Montego Bay, Jamaica, overlooking the Caribbean. The theme was, "Back to the Roots"—Pioneer Overseas Corporation roots. POC staff made sure that it was a memorable celebration.

People at this event focused a lot of attention on Suri, recognizing him as the one who had started the operation in Jamaica in the first place and established the groundwork for the success they all now enjoyed.

Tom Urban did not appear happy at this celebration, either. While still in Jamaica, he told Suri that he wanted him to change his job title and responsibilities to serve as the vice president, government affairs, instead of heading POC. Suri was startled by the suggestion. He declined the offer, and no change was made.

In retrospect, Suri and Edda agreed that the Jamaica event was probably the turning point for Suri at Pioneer.

Suri continued to operate from his Johnston office; his corporate office sat empty for the duration. He used his allotted parking space once a month when attending executive committee meetings. Suri concentrated on his continuing successes and the increased business he was capturing abroad. He had built Pioneer's overseas operation into a model organization. The compounded growth rate in after-tax profits from 1971 to 1986 was 54.2 percent. His team considered him accommodating, unbiased, fair, and loyal, always a gentleman in terms of trying to make things happen without anybody feeling disrespected in any circumstance. Playing corporate politics went against everything that had made POC such a success.

Previous Pioneer CEOs, Brown and Skidmore, felt the same way Suri did about the folly of centralizing the company operations. In 1987 Suri received the handwritten copy of a four-page letter that Bill Brown sent to Tom Urban, expressing his concerns about the changes he saw occurring at Pioneer that he believed would "impact negatively on the company's future."

Brown referred to "questionable management decisions" and reminded Urban of the "many sound principles that have served the company well." Item five in his numbered list of concerns was "Treatment of long-term productive managers." There Brown wrote several comments about Suri, including, "Surinder S. has developed one of Pioneer's fastest growing and successful operations, POC. Among Pioneer managers, he is certainly one of the most imaginative and innovative. No one can reasonably question the success of his operation. Yet instead of being awarded for his success, POC is apparently being dismantled and removed from Suri's management. In my view, persons with Suri's ability are not easy to come by. An intelligent, articulate CEO should recognize this and protect rather than harass such an employee. In my opinion, the way the president of POC has been treated is a disgrace—and that treatment will live to haunt its perpetrators."

Brown's letter ended by apologizing for offering advice not asked for and adding an assurance that his comments were made out of interest in the company.

At the same time, Suri's life at home in Urbandale had begun
to change in a number of ways—and not just because Pioneer as a
company was becoming less fun to come back to. New families kept
arriving from India, and the Sehgal kids were growing up.

Suri hadn't really been around enough to teach his sons some of the
standard father/son lessons, such as how to shave, or tie a tie. The boys
figured out those everyday things mostly on their own—or from Edda.
Suri did not participate with his children in the types of activities that
were more typical in many Iowa families. They didn't play board games
or baseball together, or go fishing or camping. Suri and Edda wanted to
find particular activities they could all share as a family. They went on
a lot of picnics, attended a few minor-league baseball games, and visited
state parks. In addition to their shorter excursions around the Midwest,
which became more frequent and often included a cousin or two, their
big summer road trips continued.

The family drove from Iowa down through the southern states
to Florida in 1982, stopping at Disney World in Orlando, and in
Homestead, Florida, to visit Suri and Edda's old friends. As was
usual on long car rides, they listened to AM radio, and tried new
foods. The children bought comic books at gas stations and sang
along to music on tape. They sang along with the entire soundtrack
of *Grease* in the car. From Florida, they traveled on to Jamaica.
There had been regular trips to Jamaica since 1970 by Suri, and
sometimes with Edda and the kids, but this was their first trip to
the island with all four kids. Kenny had become more involved in
music and now played the guitar. He introduced the family to the
music of the Kinks. Even Suri admitted to liking their music, which
was a good thing since the boys played Kinks music over and over
throughout their stay in Jamaica.

Starting in the early 1980s, the whole family learned to ski—a
perfect family winter activity during Suri's slower time at work. The first
couple of years, they went to little places in Iowa to learn how to ski on
small slopes. Then they took the big leap to a ski resort in Steamboat
Springs, Colorado, and there was no turning back. The whole family
was hooked on skiing and being in the mountains. Jay, who went along
a few times, observed that it was pure family time with no phones or
work interruptions to get in the way of the fun.

Suri and Edda supported the boys' efforts to get jobs to earn extra money for special purchases. Kenny and Ben each delivered newspapers for a couple years, the *Des Moines Register* in the morning and the *Tribune* in the afternoon. Suri drove them on Sundays, since the paper was so heavy, and also when the weather was particularly nasty. Oliver and Vicki delivered a free weekly newspaper, *The Shopper*, in the neighborhood after school. After Oliver stopped, Vicki continued on, and Edda helped her stuff *The Shopper* into plastic bags the night before delivery day.

Once kids turned sixteen, they could do seasonal work in the Pioneer cornfields. Fieldwork was something kids signed up for at school, and many did it, including Kenny and Ben. The money was considered decent for a summer job; after three or four years, workers earned raises. A lot of the kids who took these jobs did them every summer through high school and college. By that time, they could be crew leaders.

The practice of sending young people into the fields every summer had emerged with the commercial rise of hybrid corn. The first job Kenny did with Pioneer was detasseling, removing the tassels from the top of corn stalks to prevent the plant from pollinating itself. The all-day work could be grueling whether it was raining or a steamy one-hundred degrees. After two summers, Kenny decided "detasseling was for chumps," and he switched to pollinating, which he considered easier, more interesting, and it paid better.

Ben worked in the cornfields for only one summer, 1984. His job was roguing, which entailed walking up and down the rows of corn, identifying rogue stalks that did not conform to the preferred form, and chopping out each one at the roots with a spade. The job started at six in the morning. The crew of teenage boys would meet in Johnston, load up in a van, and head out to the first of several cornfields they would work that day. The boys had to wear long pants and long sleeves, even in scorching heat, to avoid getting cut by the leaves. If it rained, they walked in mud. Even if it didn't rain, they ended up drenched from the morning dew and their own sweat.

Like Kenny, Ben found the work pretty demanding, especially once the corn was over his head. The job was in no way fun and interfered far too much with his high school baseball practices.

Despite such complaints, the experience was eye-opening in some unexpected ways. Not only did the boys learn a valuable thing or two from their college-age crew leaders, but working for Pioneer gave them new insight about their father's position in the company.

During one day on the job, Kenny found out that his dad was the president of POC and a corporate vice president of the entire company.[42] Kenny didn't really know what that meant, exactly, but he saw the evidence posted on a wall in the little office where field workers assembled before and after their shifts.

His crew leader, after realizing that Kenny was Suri's son, asked him, "What are you doing here working for $3.35 an hour, starting early in the morning, all day, all summer?"

This was the first time it occurred to Kenny that he might be seen as any different than the other field workers. Suri had certainly never put his sons in any positions of advantage within the crews. They did whatever the rest of the workers did.

Kenny answered his crew leader perfectly honestly, "I want to buy a stereo!"

Around the same time, Kenny read some biographical notes in Pioneer publications that mentioned Suri's accomplishments. Though Suri had never brought these things to the attention of his family, the kids began to notice that articles about their father's achievements were sometimes covered in the local newspapers.

During the summer of 1984, when Kenny was seventeen, Suri arranged for Kenny to have a summer job in Germany, doing the same kind of fieldwork he'd done in Iowa. He went with a friend, Peter Dewald's son, Ricky, and they both worked in the fields at Buxtehude, near Hamburg. Kenny made quite a bit of cash at his job that summer, and spent so much on record albums that he had to buy an extra suitcase to bring them home.

Suri had arranged for Ben to spend that same summer with a Pioneer subsidiary in Austria, but Ben didn't want to miss summer baseball season, so he bailed out at the last minute and stayed home. When Kenny came back, talking about his overseas adventures, Ben was a little sorry he didn't go.

---

42. As part of the emerging restructuring toward centralization of Pioneer, the presidents of some Pioneer divisions were also named as vice presidents of Pioneer Hi-Bred International.

While Jay was in college, studying information technology (IT) at the University of Iowa, he lived first in a dorm in Iowa City then in an apartment with other Indian students. But he continued to bring his laundry home to Edda and raid the refrigerator in Urbandale. Once he graduated, he took a job working in the IT department at Blue Cross Blue Shield and lived with Suri and Edda for a few months before finding his own apartment.

The value of education and training in a decent, skilled profession was still a constant theme promoted by Suri to his Indian relatives. But Suri provided a mixed and somewhat lighter approach in his guidance to his own children with regard to college decisions and career tracks. They each made their own choices over time.

Kenny's first two years of college were also at the University of Iowa. He was not very happy there even though he had earned straight A's in sociology, and any other "ology," with little effort. Suri wanted him to take a course in "real" science, so Kenny agreed to try botany. He felt "tortured" by that class. With a knack for anything in the humanities and the arts, Kenny left Iowa City and enrolled at Grand View College in Des Moines to pursue a bachelor's degree in fine art. He lived in the Drake area but still had his old room in the basement of the Sehgal house in Urbandale.

Ben, who always did well in school and on tests, followed Kenny, and their cousin Jay, whom he continued to look up to like an older brother, to the University of Iowa (despite Suri's suggestion that he apply to Harvard). Ben majored in economics with minors in history and German.

Oliver and Vicki were still in high school as their older brothers were off at college. Oliver had excellent grades up until high school. The only time he remembers ever being "whacked" by Suri was on the subject of his grades during his senior year. Knowing how smart Oliver was, Suri finally had had enough of his son's poor grades and long hours playing video games. Angrily he said, "You are not studying!" and gave Oliver a smack.

Oliver reflected later, "I could have tried harder, but I was bored and didn't see the value. Dad let us, more or less, run loose and learn everything for ourselves. But when he cracked me, that smarted!"

Suri and Edda were not aware at the time of some of Oliver's more creative endeavors in high school. No mention of his pranks was ever made to Edda in the parent-teacher meetings she attended regularly. On one occasion he set off an odorless white fog bomb, causing a teacher to pull the fire alarm. Vicki was swimming in physical education class in another part of the school when everyone had to get out of the pool and go outside in the cold. Vicki stood in the winter weather soaking wet and shivering in her bathing suit. That was the only time she was displeased to be associated with her brother Oliver.

With three brothers who came before her, Vicki often saw a familiar wary look in a teacher's eyes when she was identified as yet another Sehgal. Though she wanted to create her own niche in school, she was thankful that she had her brothers to tell her what to expect in high school. Through Oliver, she met juniors and seniors, which helped a lot when she was a new freshman. Vicki studied hard in school and joined the student council, volleyball and track, the German Club, and other clubs and committees. She was still planning a career in teaching and, by her senior year, she would do practicums in the local elementary school to gain experience.[43]

---

43. Vicki began working at the local Dairy Queen the summer she was fifteen and continued to have jobs all through high school and college. Though all the Sehgals knew their college tuition and housing would be taken care of by their parents, they each always found ways to earn their own spending money.

CHAPTER 16

Sweet Poison

Pioneer Overseas Corporation (POC), led by Suri, employed more than 1,600 people worldwide in 1988. The operation had twenty-seven breeding stations around the world that were supporting seventeen subsidiaries, and numerous distributors were spread across sixty countries. The projected profit was more than 52 percent of Pioneer's total net income for the year.

Suri was called to the office of Pioneer's CEO on a Tuesday morning in early March. Sitting across the table from Urban were two attorneys. The middle chair had been left empty for Suri, who had just returned from France the previous day.

After Suri sat down, Urban asked, "Do you know why you are here?"

When Suri replied, "No," Urban quickly responded, "Your services are terminated."

Stunned, Suri asked, "Why?"

Urban replied quickly, "We don't have to give you the reason."

Suri asked again, "Why? What is the reason? I have worked here for over twenty-four years. Don't I deserve to know the reason?"

Some papers were thrust in front of him, and Suri was instructed, "Sign this settlement agreement stating that you will not compete with Pioneer for the next two years."

He was instructed to return all of his keys, the company car, Pioneer credit cards, etc., and not talk to any other employees.

When Suri refused to sign the noncompete, he was told he had until 5:00 p.m. that day to agree to the severance terms, including the noncompete clause. One attorney added in an emphatic tone, "You better think over this very generous offer."

Still stunned, Suri walked out of Urban's office and initiated a series of visits and calls to consult with a few of his closest associates. He first went to see his good friend and mentor, Si Casady, whose office was in a nearby building. Si was immediately supportive. He was glad to hear that Suri had not signed the noncompete agreement and suggested he retain an attorney. So Suri phoned the organizational consultant who had been advising him on various management issues.

Agreeing that it was a good thing Suri hadn't signed the noncompete, the consultant referred him to a highly respected attorney, Jim Gritzner.

Before heading home, Suri stopped at Bill Brown's house in Johnston, running into Skid on the way. Brown and Skid had already heard about Suri's termination from one of Pioneer's executives who was making the rounds to disclose the news in person. Suri's own POC staff had been told the news while he was still in Urban's office.

Both former CEOs of Pioneer were unhappy about this turn of events, and very sorry not to have influence over the situation.

When Suri went home and told Edda what happened, she was not at all surprised. She had watched Suri's relationship with Urban decline rapidly in recent months. She knew how unhappy her husband had become in his work environment. With her usual pragmatism, Edda had prepared a budget within the hour that showed the family's monthly needs and expenses and how they could get by for now without the income from Pioneer.

Suri met with Gritzner and his team that afternoon. Their considered advice was for Suri to sign the noncompete. They cautioned him, "You have limited resources, and this type of thing could drag on indefinitely. Pioneer is a huge company . . . the odds are against you . . . you've been offered a pretty good financial settlement," and other advice to that effect. The "generous" offer would be withdrawn at 5:00 p.m., so "consider signing it."

But Suri was firm: he would not agree to a noncompete. At the end of the consultation, a call was made on a speaker phone before the 5:00

p.m. deadline. Suri's new attorneys informed Pioneer, "No, Dr. Sehgal will not sign the noncompete agreement."

The astonishment in the voices of those in Urban's office could be heard on the phone. That Suri would leave all that cash on the table was unthinkable.

Vicki was surprised to come home from high school that day and find her dad already home. She knew something was wrong: he had never missed work. While Suri was busily digging through papers and file cabinets in his home office, Vicki asked Edda what was going on, and her mother said calmly, "Not right now. There's a lot going on, and your dad needs to find some things."

Later that evening, Edda explained to Vicki that Suri lost his job.

A day later, many were shocked to read the front-page story of Suri's firing in the *Des Moines Register*. The same attorneys who had successfully sued Garst and Thomas had worked overnight to prepare a lawsuit against Suri, which was filed in Polk County District Court in Des Moines that morning. The lawsuit alleged that Suri endeavored to take trade secrets from the company. The lawyers went directly to the media with the story—to the paper, the radio, and television.

A reporter for the *Des Moines Register* called Suri that afternoon to ask him if he had been told he was being sued. Surprised, he answered, "No." He was mystified by the accusations.

Oliver first heard about it on the radio the next morning, as did Jay.

Jay was ironing his shirt while listening to the news before work when he heard someone say that Pioneer fired the president of their overseas operations, and a description of the allegations.

After a moment, Jay realized, "They're talking about Uncle!"

He called Suri immediately.

Suri told Jay in a reassuring voice, "Just hang up, and do not worry one bit. We will sit down and talk later."

That afternoon, Jay took time off work to go over to the house and talk to Edda. She assured him as well. "Don't worry. It'll work out fine," she said calmly.

Suri joined them and explained to Jay, "Calm down; it never helps to be reactive in these situations. There is no use arguing with power. Accept the reality and go on. Everything will be fine."

Oliver had been listening to KGGO radio that morning when he heard that Pioneer had fired its vice president for endeavoring to steal trade secrets. He thought, *Are they talking about my dad?*

He went downstairs and could sense that something was wrong. Edda told him what happened and showed him the newspaper article. Kids asked him about it at school. Oliver answered that it was "bullshit."

Within days, Pioneer attorneys filed a second suit against Suri, this time in the Delhi High Court, to try to prevent him from taking control over Pioneer Seed Company Limited (PSCL), a joint venture established in India in 1977 and registered under Indian laws. PSCL had been established as a joint venture because the foreign investment laws in India at the time did not permit majority control by foreign investors. Pioneer owned 40 percent; Suri's youngest sister, Sanjogta, now living in Delhi, owned 40 percent; and Suri owned 20 percent and served as the company chairman.

From day one, the *Des Moines Register* ran the story about the legal case between Pioneer Hi-Bred International and Suri Sehgal.

Suri was very busy gathering his forces to defend his professional business reputation and protect his family's future. With little savings and suddenly deprived of all income, he was poignantly aware that Pioneer had deep pockets and plenty of institutional clout in the state.

In all his years of traveling around the world for Pioneer, the satisfaction provided by the continued growth of the business had further fueled Suri's commitment to his work. Building the company had been his focus. Now his focus turned to the work his lawyers were doing. The lawyers were impressed. Noticing that Suri was working as hard as they were, the firm provided him with an office and a desk.

Suri's attorney, Jim Gritzner, who later became a federal district court judge in Des Moines, reflected on his initial impressions of Suri and his case, saying, "As a lawyer, when you first meet someone, you're always a little suspect. We deal with a lot of people who are playing a role. Early on, it was very important for me to figure out if Suri was the good guy or the bad guy in this equation. Very quickly I found out that he was the good guy, and there were some pretty significant bad guys involved."

Gritzner described Suri as "a combination of tremendous achievement and ability, as well as humility and graciousness. Suri regarded his lawyers as lieutenants, not generals, because he remained very much in charge."

Edda observed that, during this period, Suri never got angry, impatient, or discouraged. He took control. He worked long hours with the attorneys and in his home office, documenting information. He liked his lawyers and approved of their approach. He felt sure that things would eventually be all right.

The newspaper stories were unfavorable toward Suri in the beginning, and his loyal team members took heat. Their exclusive operation had been a smooth-running enterprise, but there had always been some jealousy within the rest of the company toward the tight-knit overseas travel squad and their frequent trips to "exotic" locations. The required travel looked mostly glamorous from the outside, but as Suri and those on his team had often pointed out, "We were working. It was nothing to work sixty or more hours in a week."

One member commented, "When we were in Paris, the only time I saw the Eiffel Tower was when I flew over it!"

Only a few weeks after Pioneer filed the lawsuits against Suri, he received disturbing news about what was happening in India with PSCL. He decided he had to go there right away to see for himself. But now he had no expense account, and his international travel costs were his own to bear.

In early April, Suri arrived in India and found, to his horror, that PSCL had been stripped of its assets. Pioneer wanted to leave a shell of a company in case the verdict went Suri's way. All germplasm, cash, documents, and records had been removed; crops had been destroyed in the fields; the managing director, Roger Sawheny,[44] had been replaced; and most of the PSCL employees in all three of its locations (Hyderabad, Bangalore, and Delhi) had been hired away by PHI Biogene, a parallel company that had been swiftly set up by the Pioneer representative. The employees had been told that PSCL was "dead" as a company.

---

44. Roger, a Canadian citizen, was married to Savitri and Brij Anand's daughter, Gita. His family was living in Canada at the time.

Suri worked with his Indian lawyers over the next ten days to obtain a court ruling to regain control of the company and get back its looted assets.

Among Pioneer's tactics to thwart Suri, they filed a motion to move the PSCL litigation from India to Iowa. During the hearing, a Pioneer attorney on the stand in Polk County District Court made a derogatory statement about the Indian judicial system. He said Pioneer was seeking the change in jurisdiction because "the courts in India are very slow, and not without corruption."

Gritzner produced a transcript of the court hearing and gave it to Suri's legal advocate in India, who found a way to quote from it in front of the Indian judge, making it clear that "When this American lawyer refers to corruption, he is talking about you, Your Honor."

Suri said, "I'm pretty sure that helped us in India."

Back on solid legal footing, a new board was appointed for PSCL, and Suri was reconfirmed as chairman of the company. But only six employees remained. No one—that is, no one except Suri—believed the company had a chance to succeed.

Suri wrote personal letters to all the former employees telling them the door was still open if they wanted to return. Some did, including four people in sales. He also returned Roger Sawheny to his position as managing director.

Roger moved immediately to Hyderabad to take control of the operation there, finding the scene of devastation that had been executed on instructions from Pioneer. The research fields had been plowed under or burned. Since all research records were missing, it was almost impossible to make sense out of the destroyed field plots. Roger's immediate tasks were to secure all facilities, make an inventory of the production fields, and locate all the field records.

When litigation started in India, Suri's nephew Chander was working for Pioneer in Indonesia. After his graduation from Sophia University in Tokyo, he had attended Drake University in Des Moines to obtain his master's degree. He and his Japanese fiancée, Rumiko, were married there in 1985. Chander worked for Pioneer briefly in Des Moines before becoming the company's country manager in Indonesia. Suri's influence on Chander's life had been huge. When he heard about the situation at PSCL, Chander resigned his job and immediately rushed to India to join the six-person team in Delhi to help Suri get the company back on its feet.[45]

Soon the Dehli court again ruled in Suri's favor and ordered Pioneer to hand over the seed, the precious germplasm, they had taken. But despite the favorable rulings in India, the attacks on Suri didn't stop. The company's legal strategy was to keep Suri tied down in the US with depositions or court appearances while aggressively pushing further litigation in India.

Suri provided his fledgling team with as much technical leadership as he could in these circumstances. He was subpoenaed for depositions and was forced to return to Des Moines. While he was back in the US, Pioneer went after Roger Sawheny, to find a way to force him to leave India. Suri would travel back and forth to India a half dozen times before the end of the year.

Suri's deposition in early June 1988 lasted ten days, during which time Suri learned about the existence of a secret tape recording. The tide was about to turn.

As the facts began to emerge in the depositions, it became clear that Suri's firing was in fact the culmination of events going back to 1982 when certain corporate-level leaders at Pioneer were bent upon breaking up Pioneer Overseas Corporation and merging it with domestic operations, giving them full control of overseas business, and Suri.

The previous fall, when Suri had been explicitly told by Urban to implement the new centralized regional structure, he had become very discouraged. No longer enamored with his position at Pioneer, he had started to explore other alternatives.

The only other person from India on Suri's overseas team in Des Moines was a young man Suri had taken a special interest in helping with his career. As a close confidant, the ambitious young man was well aware of Suri's frustrations with the direction Pioneer was going. Eager to further his own standing with Pioneer's CEO, even if it meant betraying the man he often referred to as his mentor and "like a father,"

---

45. A few months later, when Chander's help was no longer needed, he returned to Japan and went into the restaurant business, eventually owning and operating several successful Indian restaurants in Tokyo.

Suri's protégé exaggerated what he thought he knew in order to convince Tom Urban that Suri was plotting to set up a competing business. Urban and his CFO planned to foil the imagined plot and finally get rid of the man who was an obstacle to their plans for Pioneer's centralization. The guns pointed at Suri in this particular instance were figurative.

At the end of February 1988, less than a week before Suri was fired, Suri's protégé had been given a tape recorder and some scripted questions designed to lure Suri into admitting his conspiracy against Pioneer.

At the time, Suri was preparing to leave for a meeting with France Maïs people. On his way to France, he stopped in Boston to meet with a group of venture capitalists. He was seeking funds for a seed venture he would either do alone or with Pioneer, a prime opportunity he had described to Pioneer's CEO and others, even saying at one point, "If you don't do it, I'll do it myself."

With a tape recorder hidden in his briefcase, the protégé had brought the papers Suri needed on the day he was leaving for Boston. But the report now contained newly inserted proprietary information about germplasm that had been added intentionally by the protégé, in spite of Suri's explicit instructions never to do such a thing. The expectation was that Suri would pass along the confidential papers to the venture capitalists in Boston, and Pioneer would then have "proof" of Suri's duplicity.

The first recording attempt was foiled when the tape machine malfunctioned. Later in the afternoon, a second attempt was made when Suri's protégé connected the tape recorder to his phone and called Suri at home just as he was about to leave for his trip. The protégé asked Suri the loaded questions as instructed, hoping he would blurt out something incriminating.

However, other than a few candid and choice disparaging remarks Suri made as he was rushing about before leaving for the airport, the tape didn't include proof of much other than his annoyance and frustration with the changes happening at Pioneer, which was no real secret.

Nevertheless, the company attorneys played the tape again and again for other employees at Pioneer, after Suri was fired, to justify the firing and to convince Suri's overseas team that he had tried to sabotage the company.

The CFO and in-house counsel kept insisting that Suri was guilty. They were so sure of it that they hired a crew with listening devices, housed them in a van, and monitored other employees as well.

The new information discovered in depositions about how Suri was set up prompted him to file a countersuit against Pioneer. The news coverage continued, and the case was now referred to as David and Goliath in the press.

When Gritzner uncovered the truth about Suri's protégé, Suri was crushed. To be stabbed in the back by someone he had taken under his wing and felt such affection for was hard to accept. He received a phone call soon after from a guy who went to college with the young protégé. Suri learned then that the "slick" man's college nickname was "Sweet Poison."

Fighting a case that promised to be protracted was an expensive proposition, so Gritzner suggested to Suri that it would be more practical for him to transfer his case to a team of lawyers willing to do the work on a contingency basis. The transfer was made, and the work continued.

The lead contingency attorney, a man named Glenn, later described what he and his colleagues saw in Suri. "Suri was clearly a very bright, well-educated man who had lots of motivation, and a wife behind him who fully supported him. That was very important. He had been groomed by the preceding chief executives of Pioneer, Bill Brown and Wayne Skidmore, particularly Brown, who had recognized that Suri had the ability to expand Pioneer's reach by going overseas and setting up international operations. By the time the lawsuit happened in 1988, Pioneer was worldwide. Their problem was that the whole overseas operation, the entire non-US operation, was loyal to Suri, because he was the guy who built it."

While Suri was in the midst of depositions, he received a call from Geert Van Brandt on behalf of a company in Belgium. He asked to visit with Suri in Des Moines to discuss what his company was doing in biotechnology. Suri agreed to the meeting.

Plant Genetic Systems (PGS), in Belgium, had been established in 1982 by a molecular biologist and professor, Marc Van Montagu, and a venture capitalist, Gerard Van Acker, for the commercialization of bio-technology inventions. Van Montagu and Jeff Schell at the University of Ghent were among the first to demonstrate a practical method (agro-bacterium mediated transformation system) for the genetic engineering of plants. Van Montagu imagined the potential for genetically modified plants to help the rural poor by increasing crop productivity and protecting people and the environment from the overuse of chemicals in agriculture.

PGS had developed the world's first genetically engineered insect-tolerant plant by inserting into tobacco a gene from the soil bacterium *Bacillus thuringiensis* (Bt) coding for a protein that is toxic to the larvae of certain crop pests. Tobacco was used for this testing because the plant was not part of the food supply.

The resulting "Bt tobacco" showed it was possible to induce plants to produce their own environmentally friendly and natural microbial insecticide. Just as antibiotics could be broad spectrum or very narrow spectrum, Bt toxins could also be very specific and targeted to specific larvae. The technology would make it possible to reduce the use of poisonous insecticide sprays in crop production.

The company had other products in the pipeline as well, including a novel plant-hybridization system, called SeedLink, and genetically engineered plants that were tolerant to the broad-spectrum herbicide glufosinate. The challenge for PGS was to tap into the vast commercial potential of their breakthrough technologies.

Van Montagu had heard about Suri from colleagues at a professional conference at Iowa State University and had suggested to the CEO of PGS that he should contact Suri.

Coming from a classical breeding background where developing new hybrids was done in the field, Suri recognized that PGS was like a biotech boutique for biologists—the "grail" itself. The promise of this cutting-edge technology offered a new world of possibilities.

At the end of the meeting, Suri asked Geert, "So, what can I do for you?"

The company wanted to hire Suri, if he was available. His long record of success in the commercial world, delivering good products for the international markets that he had been instrumental in creating, would make him an exceedingly attractive asset to PGS. The brilliant scientists of PGS worked in their labs; they had no business experience or the know-how to market their products. Suri had the exact additional skills the company needed.

As fascinating as the offer sounded, Suri said he wasn't looking for employment under the circumstances; he was just too busy.

The PGS representative replied, "Look, you may not be looking for a job, but I am looking for a boss." He pressed Suri to come to visit PGS in Ghent.

Suri agreed to a brief meeting on his trip to India in September if the company paid his fare from London to Brussels. The PGS representative agreed.

In Belgium, Suri found that PGS was indeed doing "beautiful" things in biotechnology. The people he met presented an attractive and heavy sell, insisting that Suri must join them. Suri was particularly impressed by Marc Van Montagu's interest in sustainable agriculture and his sincere desire to help developing countries with the technologies developed in his lab.

At the insistence of the company CEO, Suri canceled his evening flight, stayed for dinner, and met with more key people. But he remained undecided. He was concerned that he really didn't have time for work with PGS along with his responsibilities in India and at home and the ongoing lawsuit, which was a long way from being settled.

The PGS people responded, "If you end up with an injunction against working in the US, we don't care."

Before Suri boarded his plane home, he had agreed to consult with PGS for ninety days in the coming year.

Almost immediately upon his return home, another job inquiry came from a German seed company, KWS, which was primarily in the sugar beet seed business, but now also in corn. They wanted Suri to help them develop their business strategy in seed corn.

Attorney Gritzner explained, "Suri got those offers from two European companies, and we told him there wasn't any reason he couldn't accept them. That gave him good income. He just popped right back up, and away he went!"

Suddenly Suri was earning twice the income he had in the past. Money was no longer a problem.

The public drama of the lawsuits would last for another fifteen months. The snare that the company leadership had laid for Suri would eventually recoil back at them.

Gritzner explained, "In the beginning, it looked as if Pioneer had some dirt on Suri. But, as it turns out, they were wrong. They underestimated Suri significantly. Before he met with the venture capitalists in Cambridge, he had looked over the document, noticed the added information, and quite appropriately withheld it from them."

In addition, Suri's personal reputation within his team and circle of work associates was beyond reproach. Although there was a lot of kinship in POC, and some well-deserved celebrations, Suri had always kept his business and his social life separate. Despite frequent invitations, he had never even joined his team when they went for drinks on Fridays after work.

Nobody could demonize Suri, be it for character, finances, or ethics. There was no indiscretion to pin on him. He had always been focused on business, and work came first.

A weakness in Pioneer's argument had stood out immediately. No one could offer an explanation as to exactly how Suri planned to do the things they accused him of trying to do. The seed stocks were locked up, and Suri had no access to them. Once Urban acknowledged in his deposition that he had handed the tape recorder to Suri's protégé, and it became known that Pioneer's lawyers had provided the questions, the tide turned in Suri's favor. [46]

That the protégé inserted the confidential information into the report and gave the report to Suri—in the hope that he would pass it on to the venture capitalists—was "pure entrapment." And the fact that they fired Suri first, then demanded that he sign a covenant not to compete, was an antitrust violation (restraint of trade). In addition, Pioneer's leaders had been telling other companies not to employ Suri—another restraint of trade violation and cause of action, along with breach of contract and tortious interference.

After the series of depositions of key players from September to December 1988, and especially the "bombshell discovery" of the tape

---

46. In a conversation with a colleague from Australia who was in Des Moines on a routine business visit, Suri's protégé had openly divulged his role in trying to entrap Suri.

recording, Suri's legal team spent the next few months preparing their case against Pioneer, which was scheduled to go to trial on August 31, 1989. Before that time, there was plenty of haggling and attempts at settlement.

When the judge ruled on the pretrial motions, the issue was that Suri was being sued for endeavoring to take trade secrets. But no one could articulate what they thought he was trying to take, except as mentioned in the protégé's testimony, which had come across "weird." The judge pointed out that, because Suri didn't physically take anything, the case was an "empty shell."

During his remarks, the judge added that one of Pioneer's attorneys had behaved in a shameful fashion in a meeting with Suri's attorneys. He had made threats along the lines of, "When we're done with Suri Sehgal, the only thing he'll be able to plant are flowers in his backyard. That's the only genetics he'll be concerned with!"

Then there was the issue of the tape recorder and the questions.

After the judge's ruling, just before the trial was to start, the judge again asked the parties if they would be willing to confer about settling the case. This time Suri agreed.

The unseemly drama ended in a settlement that day between Pioneer and Suri. All charges were dropped. Suri was reinstated as vice president of Pioneer Hi-Bred International, Inc., and as president of Pioneer Overseas Corporation.

He immediately resigned.

Suri had two different compensation options in settlement discussions. He was offered cash, or a lesser sum plus Pioneer's shares in the company in India. Included in that offer was half of the germplasm developed by PSCL in India—some of the very seed stock Suri had been accused of taking in the first place.

In the end, Bill Brown's prophecy came true. The attacks on Suri ended up haunting the perpetrators. A few company executives were just happy to have Suri gone. After all, he had already put in place the highly lucrative worldwide business for Pioneer, and now he would no longer be an obstacle to centralization.

However, the India settlement continued to be an ugly fight. Suri's lawyers had to go to the judge at least a dozen times to get documents turned over properly. Pioneer attorneys were fined a couple of times by the judge in the US, all of which was described in the *Daily Business Record*.

The newspaper coverage turned around in Suri's favor now, but the articles were no longer on the front page.

CHAPTER 17

Global Initiatives

As the lawsuit went on over seventeen months, life at home in Urbandale had continued on as usual. In fact, life went on with a good deal less stress. Suri no longer felt the growing sense of unease he endured in the period leading up to his departure from Pioneer. Though Suri's colleagues at Pioneer had been told not to have any contact with him, many of his associates stopped by the house individually and in small groups to share their support, enjoy a drink, and express their good wishes.

There was little discussion or concern about the legal case at home. Vicki noticed that her father still traveled a lot, and when he was home, he got up early every morning and went somewhere, though she wasn't sure where. When the stories were in the paper, her friends sometimes asked how things were going. There wasn't much to say.

When Kenny came home from college for spring break and read the articles that disclosed the salaries involved—he learned that his father made "a relatively large amount by anyone's standards."

One of his friends had joked, "Well, Kenny, when are you buying me that Corvette?"

The Sehgal kids knew their parents had both grown up during the war and their families had struggled to survive. They knew that Edda never threw away a scrap of food. She put any leftover food in the fridge, thinking it would be eaten later (and it was). The family didn't have cable TV or a big stereo. The boys drove the same used VW Rabbit that Edda had driven—and before that the Saab that often broke down in the snow.

The settlement with Pioneer did not make a difference with any of these situations.

Ben observed about that period, "Our lifestyle didn't really change. But I think that was the first time I got the sense that maybe we had assets I was unaware of before."

Suri and Edda did not believe in borrowing money. The only time they did so was when they bought the house in Urbandale, and they paid off that mortgage as soon as they could. They never paid interest on credit cards, because they never purchased anything they couldn't afford. When Suri first lost his job, and his lawyer had said he was eligible for unemployment, Suri had answered, "What? Nothing doing!"

Edda commented, "If you live modestly, you don't have problems."

The only expensive "hobby" they had was making sure their extended family was well settled and their kids were getting good educations. When any financial help was needed from Suri and Edda, it was given generously with no strings attached.

Suri and Edda had established a corporation, called Biogenetic Technologies (BGT), soon after the lawsuit was filed in 1988 to protect the family interest in PSCL. After the settlement in 1989, they established another corporation under the name of Global Technologies primarily for Suri's consulting work in Belgium and Germany. Former Pioneer colleague Peter Dewald joined Global and they hired a part-time secretary to work in the Des Moines office. Edda worked there as well and did the accounting and bookkeeping for both companies, just as she did for the family.

With the legal issues out of the way, Suri was finally able to again focus more attention on PSCL in India. Roger Sawheny had worked hard to stabilize the Hyderabad operation. But Pioneer had successfully kept Roger from any further employment in India, and he had returned to Canada. Sanjogta's husband, S.K. Kapoor (S.K.), was able to step in, on a part-time basis at first (he was working for an oil company prior to that), to take Roger's place and help set up an organizational structure for the company while Suri's lawsuit was continuing.[47]

A young man from Kashmir named Puneet had an office next to the PSCL corporate office in Delhi. Working as Pioneer's travel agent, he was in close proximity to the goings-on before, during, and after Suri left Pioneer.

The first time Puneet met Suri, while he was still with Pioneer, was after the local Pioneer manager told him that a "VIP" was coming to Delhi. Puneet was instructed to get a Mercedes Benz for Suri to drive, book a suite at the Hyatt, and find a five-star restaurant.

Puneet recalled, "But here comes Suri. This 'big' man surprised me. I thought he was going to be a big man with all the airs. But he ordered food from the local restaurant, fast food, takeout. I wouldn't eat there normally myself! He always stayed in simple places. If the air-conditioning didn't work, that was no problem. If the shower in the guesthouse didn't work, that also was no problem. He worked in one of the office cubicles like any other employee."

In those days, the airline schedules were in two thick volumes, about 5,000 pages each. Puneet was impressed that Suri had his schedule down to a science. He was able to leaf through those cumbersome books and figure out his flights.

As soon as the lawsuit was filed against Suri, Puneet was asked by the Pioneer manager to supply five buses to bring all PSCL employees from Hyderabad, Bangalore, and Delhi to a five-star hotel in Delhi on the same day. He was told that the company was "breaking away."

Without any advance notice, the employees were given a choice between staying with PSCL or joining the new company Pioneer had just created, PHI Biogene. Everyone immediately had to sign a document saying whether they would stay with Pioneer or leave. The six people who decided to stay were told, "Okay, if you want the office, you can occupy it, but there is no system, no technology, and no cash. You may or may not get a salary."

---

47. S.K. became a full-time employee after the lawsuit was settled. He steered the company through its growth period until its divestment and then stayed on for three more years. Roger and his family eventually moved to Des Moines where he accepted a job offer in an engineering firm.

Within twenty-four hours, the place was ransacked; germplasm, documents, and all the cash had been taken, and the offices were empty. Puneet tried to reach someone in authority at Pioneer to get the money they owed him, but no one would take his call.

Once the lawsuit was settled, and Suri officially owned all shares in PSCL, he renamed the company Proagro Seed Company Limited. Now, in addition to his consulting work in Belgium and Germany, he embarked on the development of a new seed business in India. He was ready to focus on rebuilding the company he had started when PSCL was a joint venture with Pioneer.

From the time he began working for Pioneer in Jamaica in 1964, Suri had created many new companies, and he still had all the international contacts and friendships. But this company, his company, had literally been trashed; rebuilding had to start from scratch.

As the process began, Puneet, still working next door to Proagro's Delhi offices, enjoyed watching how the man who had been the global head of a fortune 500 company started a "mommy-daddy shop" with only a few employees. "Suri took control and grew the company step by step. He was a survivor—lucid, low-key, and always at his task."

Suri was sometimes frustrated by the bureaucratic nature of anything he tried to accomplish in India. He was far more comfortable with Western ways of asking straight questions and getting straight answers. In India, he had to weave around a few corners to get things done. Puneet became a local ally, offering to help with some of the bureaucratic processes, getting licenses, and explaining how to navigate the red tape and restrictions common in the License Raj.[48]

Puneet noticed that when anyone tried to play games with Suri, he would not be bullied. No matter how frustrated he might be or how numerous the hurdles, he carried forward with his vision and pursued it with vigor. Suri knew that Proagro could be built into a sustainable business.

Though Suri started with so few employees, he considered his team a group of very good people. His legal settlement had sent shock waves

---

48. A term referring to the licenses and regulations required to establish businesses in India from 1947 until 1991 when reforms were enacted.

through PHI Biogene. Many ex-employees of PSCL now wanted to come back, but Suri decided not to take any of them back—particularly now that he had a better understanding of the way the business had been handled during the lawsuit.

Proagro's company culture did not emerge overnight. As he always had, Suri applied sound business principles to the growing operation, and as chairman, he led through example. He listened to people and was fair-minded and principled. To build a great organization, he understood the value of strengthening the company's human capital. He identified good, honest people to work with and gave them the opportunities to learn and grow with the company.

Talented people came to work at Proagro, including young scientists and others who found themselves working in a company that felt like family and was not driven strictly by the bottom line; it was important to also create social benefit to farmers.

Suri recalled, "There was a lot of handholding in the beginning. They were youngsters!"

One of the first new people was Suri's nephew, Raman Sehgal. Jay's brother Raman had earned a degree in business at the Ateneo de Manila University where he met his wife Loida. He had first gone to the Philippines to train with Pioneer in Manila.

Raman shared Suri's business acumen, his values and work ethic, along with his dedication to excellence. Raman had known his grandparents Shahji and Shila well, and he had been a witness to their lives after Suri left for America.

Suri had supported Raman's education in the Philippines and encouraged him to develop his skills. When he had money left over after earning his MBA, he tried to return the excess amount to Suri. But Suri would not take it, instead saying, "Do something else!" So Raman went on to obtain a diploma in computer data analysis.

In 1989 Raman was working in systems consulting for a trading house, and Loida was employed at Asian Development Bank on an official mission to Papua New Guinea. They were planning to go to Australia to explore immigration opportunities.

Suri called Raman in the Philippines in August, as the court case was coming to an end, to invite him to come instead to India to work as a systems consultant with Proagro. Raman jumped at the opportunity and took the next flight to India. In the next six months, he took four trips to India, helping to set up data analysis software for Proagro's research and development group. When that project was completed in March, Suri knew his nephew was ready to do more. Suri wanted Raman to take charge of Proagro's operations at Hyderabad; the new job began in April 1990.

Important changes had already begun to occur as early as 1988 that furthered Suri's efforts with Proagro and helped the Indian economy as well. First of all, the rains came! With the end to a severe drought that had plagued the seed industry for several years, there was a big demand for quality seed. At the same time, the seed industry, which had always been a monopoly of the government, was now being privatized, and Indian investment laws were relaxed. Foreign investments in many industries, including agriculture, were now encouraged.

In the first couple of years of Proagro, there was a good demand for seed, but little seed was available due to the complications with the lawsuit and the destruction to the fields. Taking on many important capital projects so quickly resulted in a substantial shortage of working capital.

Creditors and growers had to be handled with patience and fairness for their trust to remain intact. Suri never drew a salary, and on two occasions provided personal loans to Proagro, keeping the company on track until bank loans came through. The high grain prices and the willingness of farmers to buy seed on a cash basis improved Proagro's cash flow.

Suri and Edda now worked as a team to build the company. Edda attended all Proagro business meetings from 1990 onward. She was relied on to provide wise counsel at critical junctures. She agreed with Suri's leadership approach, particularly with regard to taking care of people, saying, "People make the organization. We need to guide them properly and understand them."

Operations remained flexible, based on honesty and trust, commitment to customer service, and an on-time delivery policy. The staff was brave and frank in their dealings; unintentional mistakes were written off as learning experiences. There was no tolerance for dishonesty of any sort, and some people were let go for that reason.

Establishing a reputation as a credible and trustworthy enterprise was paramount. Profit or no profit, Suri insisted firmly that they would not bribe anyone, and they would keep all their promises. Though money was tight, Suri would not allow anyone to postdate a check to pay a bill. When there was a corn failure in Bihar, local farmers demanded a refund of the money. Proagro had sold the seed, but the corn didn't grow due to drought in the area. Nevertheless, the money was refunded to the farmers without delay. Other seed companies simply did not do that sort of thing. The word spread, and it had a multiplier effect. Dealers started rating Proagro above every other seed company in India.

At a strategic planning retreat in 1991 with all key staff members held in Katmandu, Nepal, a collective decision was made to limit the business to four crops: corn, millet, sorghum, and sunflower. The goal was to become number one in sales in each. At the meeting, Suri gave the group a revenue target of one billion rupees to achieve in five years. His charge to his team was to take care of customers, keep a low profile (don't be flashy), do the research, access technology where possible, and utilize global linkages.

The goal was to build something to last. Suri, Edda, and their team were convinced that they were running a business that was doing something good for society. That, not the bottom line, was the driving force behind their commitment.

Suri invited his contacts from abroad to visit. Business was prospering, and Proagro now had beautifully maintained facilities. Suri and Edda knew that when their international guests returned home with good impressions, Proagro's reputation would develop even faster.

For Suri, there was no compromise when it came to investment to further the business. The telecommunication industry in India was still poorly developed, and it was a challenge even to make a long-distance phone call. Proagro, ahead of the industry curve, had established an IT department and computerized its offices and operations with the help of Suri's nephew Jay.

Jay was still working in the IT department at Blue Cross Blue Shield in Des Moines when Suri offered him the opportunity in 1991 to move back to India and work at Proagro. He paid off Jay's college loans, made the transition as easy as possible for Jay, and gave him a free hand in setting up Proagro's IT systems.

Jay recalled, "I anticipated a difficult adjustment after being in the US for so many years, but the openness of the ambitious Proagro team drew me in and kept me working at my best. The company had very strong fundamentals. The atmosphere was highly decentralized and conducive to work. Everyone was energetic and helpful, and everyone wanted the company to succeed. We had ambitious goals. I learned how important it was for me to understand hybrids. Although my job was IT systems, Suri had foreseen that I needed to understand the relationship between IT and other functions of the company: production, packaging, invoicing, and moving seed to the dealers."

Even though finances were very tight and computers were expensive, Proagro imported Compaq machines and deployed the best technology. The goal was to computerize and automate every scenario in the sale and movement of seed.

Computerization of processes was something new, and Jay found ways to make the process lean and efficient. He spent time with managers to explain what they were trying to get done, and he handled any trouble that arose in unexpected situations. Within a year, the basic core warehouse operations were automated.

By 1992 dedicated breeders came to Proagro from ICRISAT (International Crops Research Institute for the Semi-Arid Tropics), and from agricultural universities. They had access to the germplasm from public institutions and foundation seed companies in the US, and they capitalized on linkages with international agricultural research centers (CIMMYT in Mexico, IRRI in the Philippines, and IITA in Nigeria) and national agricultural research organizations in India, Thailand, and elsewhere. Crop specialists from abroad regularly visited Proagro to review and help improve research effectiveness, providing opportunities for Proagro breeders to learn from the best. Most of these visitors had had close working relationships with Suri in the past. They participated in the internal meetings and shared their views with the staff, boosting knowledge, building confidence in the breeding teams, and providing input to management.

Raman said about that period, "There were always challenges, as well as the pleasure of finding good solutions. It was one of the most wonderful times of my life.[49] Suri and Edda came at least twice a year. They stayed with us sometimes, and this gave me an opportunity to spend a lot of time with them, talking about the business. Their visits were always very inspiring. Both of them participated in the strategy meetings with the core team. I learned a lot."

The company never declared dividends; its earnings were plowed back into the business. Proagro reinvested profits into expanding research, creating fixed assets to accommodate growth, and building human capital. Every employee received a bonus at the end of the year that was equivalent to a month's pay. The healthcare of employees and their families was a company priority. Not surprisingly, Proagro employees were highly motivated and loyal to the organization.

Proagro was a business with a human heart. Employee feedback was encouraged on an everyday basis, and good ideas were incorporated. The Proagro team was given authority down the line. Marketing people were able to make decisions in the field, without going all the way up the hierarchy for authorization. Some people could not handle it, but most thrived.

Suri's five-year revenue target was met, Proagro was number one or two in all four of its crops, and the entire operation was completely automated. For the first time in the Indian seed industry, a computerized process tracked the movement of seed across functions and locations within a system that provided excellent quality control and monitoring.

Proagro was the first to give incentives to seed dealers to travel abroad, people who had never before left their villages. Mailings were sent out to dealers, offering data programs and other perks that no one else had offered. Puneet recalled getting travel visas for the dealers, "farmers who did not even know to wear pants or shoes. They wore rubber slippers, which had been through a couple of repairs. Those were their customers. Years later, those same people drove fancy cars!"

Suri's commitment to rural farmers was obvious to his colleagues and to the farmers. He changed the lives of employees and villagers

---

49. Raman's wife, Loida, eventually resigned from her job at Asian Development Bank and joined Raman at Hyderabad in 1992. Their daughter Alexandria was born in 1993.

by offering high-yielding hybrid corn to India. Many farmers came to know him by name. He could often be found in the fields, talking to field assistants and others, asking them what they were doing and how he could help.

By focusing on research and high-performing products, Proagro was becoming a star performer and one of the largest hybrid seed companies in India's private sector. The construction of an ultra-modern seed-processing plant at Toopran, Andhra Pradesh, reinforced Proagro's image as a "total quality" company with superior products and excellent people.

A mechanical engineer named Raju was responsible for plant instal-lation and construction of the Toopran seed plant. Peter Dewald, who had worked with Suri's POC team, was brought in to do the design. Jay had the network wiring done during the plant's construction. The warehouse module became the core for the future structure upon which other information systems were attached, including a system for quality assurance to determine if seed could be shipped out or not.

The full plant was built in stages over three years. For installation of the expensive equipment, experts advised Raman to hire external professionals. But Raju felt that Proagro people could handle the installation and that the people who would be running the plant would benefit from the training. Like Suri, Raman trusted his people; he gave Raju the go-ahead. The equipment supplier from Denmark sent an engineer to supervise, and Proagro people did the entire installation quickly and to everyone's satisfaction.

Until then, the Indian seed industry did not appreciate that ware-house maintenance or seed processing (cleaning, treating, and bagging) were factors that improved seed quality. The Toopran plant utilized techniques like vacuum dewatering to obtain smooth concrete flooring that seed would not stick to. This attention to detail and emphasis on best practices separated Proagro from its competition.

When the Toopran plant was completed in 1995 it was the largest seed plant in Asia. Quality assurance was simple, fast, and efficient. Proagro was also the first seed company to have a mainframe accounting package, integrating data among all the branches, including Hyderabad.

Some doubted that a plant as large as the Toopran facility was needed. When Suri spoke at the inauguration of the new plant, he answered those concerns, saying, "This plant is big because you have to think about the future!"

Sure enough, the plant was utilized to its full capacity—eventually processing up to 20,000 tons of seed within one season. Even though hybrid seed was still a novelty in India when Suri started Proagro, Proagro hybrids were now planted by millions of small farmers. The company became a trendsetter. Entire villages were contracted by Proagro and other companies to multiply hybrid seed in isolation to maintain genetic purity. In these "seed villages" in Andhra Pradesh and Karnataka states, incomes rose sharply and poverty decreased.

Suri's contacts and friends from all over the world visited Proagro. People with know-how in the business came from the US, Japan, Australia, Belgium, and Germany. Success stimulated investment in the hybrid seed business by entrepreneurs and multinationals, and major global players entered the Indian market.

During the same years that Suri was busy with Proagro, his consulting work had kept him traveling in Europe. He combined some of his regular trips to PGS in Belgium with a few days in Germany. The work at PGS had evolved beyond consulting. Suri had walked into a chaotic operation with brilliant PhDs doing their own pet projects, without any notion of value recovery. These scientists were the cream of the crop from Ghent and Brussels, and Suri enjoyed working again with colleagues who matched his own high level of scientific knowledge.

In early 1990 the CEO, Walter DeLogi, asked Suri to become the chief operating officer of PGS and offered him a five-year contract. Suri asked for a three-year contract. Sure that Proagro was about to blossom, Suri did not want to be tied down.

DeLogi countered with, "You can leave when you've trained Jan Leemans as your successor."

Suri was pleased to agree to that, because Leemans, like Van Montagu, was a talented and distinguished scientist and a pleasure to work with. Leemans was skilled in explaining complicated science and intellectual-property rights issues, which Suri was interested in.

Suri's initial work at PGS involved streamlining the organization and inserting business discipline. He worked closely with Leemans in building the team and making sure it was functioning smoothly, with clear-cut goals, as well as accountability and responsibility.

He recommended stopping several ongoing research projects for which value recovery was questionable. He suggested that the researchers work on certain traits in select crops, and convinced them to focus on crops with high commercial value, such as hybrid canola. The company culture shifted from "biotech boutique" to being product-driven. Some key scientists became business managers. Canola, corn, and vegetable seeds became the strategic crops.

Suri rented an apartment in the beautiful medieval city of Ghent for his three-week stays. Edda accompanied him sometimes, using the opportunity to also travel to Germany to see her family. Vicki, now in college, spent a semester at Trinity College in Carmarthan, Wales. Oliver and Edda visited her, and together they went to London and to see Suri in Ghent. When they visited Suri's office there, Oliver was impressed by the obvious respect that Suri was shown by the people he worked with.

Oliver loved the refinement and panache of the people he met in Europe. He frequented the bars and talked to everyone he met. Vicki enjoyed walking down each little picturesque street in Ghent until she got completely lost. In such a small city, she easily found her way back, marveling at the beautiful facades and windows.

Suri was in Belgium on March 8, 1991, when he received a message from Edda that Bill Brown had died unexpectedly. There was no immediate funeral; Brown had donated his body to the University of Iowa for medical research. Suri was profoundly distraught to lose his mentor and dear friend. Dr. Brown had been a trusted colleague, traveling companion, and wise counselor from the time Suri graduated from Harvard, as well as (along with Alice) a steady source of warmth, affection, and support to Edda and the kids.

When Suri returned from Belgium, he went straight from the airport to see Alice Brown, just as he had always stopped in to see the Browns before and after each of his trips to Belgium. Bill Brown had been enthusiastic and supportive of Suri's venture into the biotech world. They had many discussions about the potential for biotechnology to perhaps do for the biological sciences what information technology had done in other disciplines. Both scientists knew, however, that biological materials are far more complex than digital bits. Even when gene transfer is accomplished, there are many other hurdles to be overcome before a product could become commercially viable. The biotech field was full of possibility.

In a recent interview, Brown had said, "This is the most exciting time in the history of botany or biology. The reason is molecular genetics. Through the techniques of genetic engineering, we can now move genes from one organism to another without carrying extraneous materials."[50]

With heavy hearts, Suri and Edda attended Brown's memorial services in Washington DC, where he had been so involved with the National Academy of Sciences and the Department of Agriculture. The whole family attended the Quaker memorial service in Des Moines. People stood and shared their memories; most of the speakers were Brown's family members.

Edda remembered sadly watching Lila and Cara, the Browns' granddaughters, at the service. Oliver remembered his father's strong emotions on the day Suri would recall as one of the saddest of his life.

Suri described his feelings, "I broke down. It was so touching, as people were talking. Cara sang Dr. Brown's favorite song, 'Wind in My Sails.'"

While at PGS, Suri had become increasingly concerned with intellectual-property rights issues. He saw that patents by multinationals were locking up technology in the private sector, and feared that such a trend, if allowed to continue, would be detrimental to the interests of small farmers in the developing world. He contended that if high-yielding varieties of wheat and rice in the mid-1960s and 1970s had been protected under intellectual-property laws by the private sector, the Green Revolution would not have occurred. Though Suri was running

---

50. *The Story of Corn* by Betty Fussell. Knopf, 1992, 75.

a private company, he vigorously promoted making biotech traits freely available to the Third World through the CGIAR centers.[51]

Suri thought that certain crops in India were good candidates to benefit from technologies developed at PGS—specifically, the SeedLink hybridization system, transgenic Bt technology for insect resistance, and herbicide resistance technologies. Although there were considerable regulatory hurdles in India, Suri believed these biotech tools would help small farmers. As usual, he did not hesitate to invest in new technologies. Proagro-PGS, a joint venture established in 1994, was the first private-sector company in India to establish a state-of-the-art biotechnology laboratory. Soon after completion it received government recognition for in-house research in genetic transformation and molecular marker technologies. Proagro-PGS incorporated the PGS hybridization system into the Indian mustard species and its Bt technology into vegetable seeds adapted to Indian climate conditions.

When hybrid rice germplasm became available in the public domain from the International Rice Research Institute, Suri pursued a partnership with several leading Japanese companies (Marubeni, Japan Tobacco, Okura) with the help of his friend and former colleague, Gensuke Tokoro, who had left Pioneer. Hybrid Rice International (HRI), a joint venture established in 1995, was the first company in India totally dedicated to hybrid rice, including breeding, seed production, and commercialization. HRI soon became and remained a market leader.

Proagro now had seven or eight satellite stations in and around the company hub in Hyderabad, with other locations in Bangalore and Aurangabad. Each site specialized in one or two crops with dedicated teams for each. Annual growth was 30 percent or more, and the company was number one in India with an excellent reputation.

At the human level, employees were pleased to work in a place where people were valued. Staff turnover was minimal. Customers also believed in Proagro. Whenever a new hybrid was released, they were

---

51. The global research partnership called CGIAR Consortium of International Research Centers coordinates and funds crop breeding research in an effort to reduce rural poverty, increase food security, improve human health and nutrition, and sustainably manage natural resources throughout the world. http://www.cgiar.org/.

assured that it was a good product that had been thoroughly tested for its performance and yield stability. The success of Proagro was the envy of the industry.

Inevitably, by the mid-1990s, both PGS and Proagro attracted the attention of bigger players who wanted to combine crop protection with crop production. Monsanto, the giant in the chemical/agbiotech arena, reached around the world to aggressively acquire seed and biotech companies with valuable germplasm and/or patents. Other chemical giants, Syngenta, DuPont, and Dow, pursued similar strategies.

Two chemical companies in Germany, Hoechst and Schering, had merged their crop-protection units in 1994 to form AgrEvo,[52] a joint venture that made a strategic move to become a serious player in the new global biotech industry. One of their first acquisitions was Plant Genetic Systems of Belgium, giving AgrEvo access to PGS's broad portfolio of traits and enabling technologies.

After the acquisition of PGS in 1996, AgrEvo invited Suri to stay with the company as a senior advisor to facilitate the integration of PGS and AgrEvo as well as to help AgrEvo with their seed business strategy.[53] As Suri described it, "PGS had extensive knowledge of plant biotechnology and a good portfolio of patents. The technologies still had to be converted into products that would bring in revenue. There was a clear realization at the top that seed is the primary delivery system for all biotech traits and technologies, therefore, forward integration into the seed business was a logical way to recover value."

AgrEvo was keenly aware of Monsanto's aggressive moves in the market. They feared that their agricultural chemical business might suffer without further investments in biotechnology and forward integration. Monsanto had already acquired several large seed companies and was sending the valuation of these companies to unheard-of levels.[54]

Since acquisitions of seed companies in the US were now prohibitively expensive, AgrEvo scouted for companies outside the US, hoping

---

52. The German company's formal name became Hoechst Schering AgrEvo GmbH, AgrEvo for short.
53. Suri stayed on as a senior advisor for the next four years.
54. Monsanto spent about ten billion dollars acquiring seed companies globally in the 1990s. Dupont had acquired all shares of Pioneer Hi-Bred by 1999 and changed its name to DuPont Pioneer in 2012. http://www.corporatewatch.org.uk/content/dupont-overview.

to acquire seed businesses in Western and Eastern Europe, Argentina, Brazil, India, Australia, and anywhere else an opportunity arose. Their strategy was to create a critical mass of germplasm and establish market presence abroad, and then make an entry into the US. Proagro was a prime target company in India because of its dominance.

By 1998 the company that started with a handful of people had more than 650 employees. By focusing on developing high-quality germplasm, establishing international linkages to access new technology, building a legacy of professionals, and investing in modern infrastructure, Proagro was now number one in hybrid corn and an important player in sorghum and millet hybrid seeds. The company was a model business organization with excellent products and the most up-to-date facilities. Proagro also had an exceptional team and a company culture that kept people motivated and unified in purpose.

When the offer for capital participation from AgrEvo came, Suri had to think about it. He had been getting offers from foreign investors, who wanted to buy equity in the company. But Proagro did not need cash for expansion. There was money in the bank. However, the offer from AgrEvo was attractive!

Negotiations went on for about a year. The final discussions, lasting three or four days, took place in Germany between AgrEvo representatives, Suri, and his company's Des Moines attorneys. The details were kept very quiet throughout the process. The final sale amount was noted in a confidentiality agreement, the contents of which were known by no more than four people in the company.

Once the sale was finalized on August 31, 1998, Suri flew to India to meet with his Proagro team face to face and explain his rationale for selling the company. He described the situation, explaining that large companies were acquiring smaller companies and that new technology was coming and would be needed; though Proagro was in sound financial shape, the company did not have the kind of money needed for the investment required to be in the big leagues.

"Suri protected our interests. He made sure that no one would lose his job," Raman said.

Beyond that, Suri's handling of the sale of Proagro was unique in the industry in that he shared his personal wealth from the sale with all 650 employees, from top to bottom. The Hyderabad office was fully involved in the divestment process. All processes and systems procedures were followed to maintain transparency and consistency.

The Proagro team was well rewarded in the form of what Suri called "thank-you money." The top-tier leaders each received a substantial sum, making them quite wealthy. Every member of the management team was given a sizeable check at an informal dinner gathering. An additional check came later. The remaining employees—from janitors to drivers—were flabbergasted to receive their generous rewards. Many people now had the opportunity to construct or finish their homes and/or invest in other business ventures.

Suri's well-trained team members were sought after by other companies, but most of the employees stayed with Proagro after the sale. Eventually, some went on to lead other seed companies in India or follow other opportunities. Through the years, former Proagro employees continued to pay heartfelt homage to Suri for his integrity, generosity, foresight, and business acumen.

Back in the US, in November 1998, the time had come for Suri to consult with his family.

CHAPTER 18

Giving Back

Suri and Edda called together their immediate family in November 1998 to share the news about the sale of Proagro. By this time their four children were living in different parts of the country. The invitation to Kenny, Ben and his wife Maureen, Oliver, and Vicki was to meet in Chicago to talk "about the future." No other details were offered.

Kenny now lived in New York City. After finishing his bachelor's degree in fine art at Grand View College in Des Moines, he had spent a couple of years painting in Chicago before getting his master's degree in fine art at New York University. He was now part of a rock band, playing guitar and singing at venues in and around New York City.

Ben and Maureen were living in an apartment in Evanston, Illinois, where Ben was completing a PhD in biophysical chemistry at Northwestern, and Maureen was teaching high school English. Ben had met Maureen at the University of Iowa and, after a couple of post-college years in Chicago, moved with her to Columbus, Ohio, where she earned her master's in education and English literature at Ohio State. Ben earned another degree while there, this time in chemistry, having decided that "real" science was his true calling. The two were married in 1994 in Maureen's hometown of Park Ridge, Illinois, and moved to Evanston in 1996.

Oliver had completed his degree in biology at Drake in Des Moines in 1993 and spent three months in Germany working in sugar beet and corn fields and visiting his grandparents, Heinz and Margarete. His entrepreneurial ventures had eventually taken him to the West Coast. He was working for a large computer software company in San Francisco when Suri and Edda called their kids together for the family meeting.

Vicki had earned a liberal arts degree from Central College, a small private school in Pella, Iowa, that was part of a Dutch Reformed community. She was teaching at a school that served low-income kids, finishing her master's degree in education at Drake, and still living in the family home in Des Moines. She found her job inspiring and fun. She had already been dating her future husband, Ryan Clutter, for several years.

In addition to handling the books for the family business, and going to India for Proagro meetings, Edda had gone to Europe frequently through the nineties when Suri was with PGS in Belgium. Those trips allowed her to help out her parents, who were experiencing health issues. She would take an all-day train from Ghent to see them. Though weakened by a lung condition, Heinz was still able to drive and picked up Edda at the station. Each visit had its own special memory. Besides seeing friends and her brother and sister, Edda helped Margarete tidy the house from top to bottom. Though her vision was declining rapidly by then, Margarete still felt strongly about cleanliness.

When Heinz died suddenly from a heart attack in October of 1996, Vicki was able to take time from her first teaching job to attend his funeral with Edda. Vicki helped pick flowers beforehand. She recalled, "I was surprised that Opa wasn't dressed in his best suit, as people were typically buried in America. He was in a christening gown. He looked so natural and peaceful."

In December that same year, Suri signed a consulting agreement with AgrEvo that required a move to Frankfurt, Germany. AgrEvo provided Suri and Edda with a house in Königstein, a picturesque small town on the edge of the Frankfurt Rhine Main Region. From there, Edda could more easily continue to help Margarete.[55]

She and Suri flew to Chicago from Germany for their family meeting.

---

55. Margarete died in 2002 and Vicki accompanied Edda to the funeral.

Happy to gather together in a hotel conference room near the Chicago airport, the family listened as Suri explained the choices and judgments made with regard to the sale of Proagro. He shared his thought process and the negotiations over the previous year, the decisions he weighed, including his consideration of whether anyone in the family might want to inherit or run the company. He described the offers he had received and the final sale that had been completed. He had everyone's full attention.

Ben later referred to Suri's news that day as "a bombshell."

Kenny added, "My jaw dropped. Literally, my jaw dropped."

Suri went on to make clear his objective to assure the financial security and well-being of his family, but that he and Edda had other plans for the majority of the money. He stressed that the family's input was important.

Just as service to others was an unspoken but implied imperative in Suri's family during his childhood in India—from his grandparents and his parents, and within the context of their Gandhian ideals—Suri and Edda's children were in full support and not at all surprised when Suri said that he and Edda wanted to use the bulk of their money to create a foundation dedicated to helping the rural poor in India.

Even though the thrust of this major philanthropic endeavor would be pointed toward India, Suri and Edda wanted the family to feel free to identify opportunities for meaningful smaller projects in their own communities, wherever help was needed. Everyone's ideas and involvement were welcome and encouraged. The family talked together about the responsibilities that come with philanthropy and excitedly began planning for the future.

Every family member was on board fully with the foundation plan. Ben, Oliver, and Vicki had each spent some time in India during the Proagro years, and all the Sehgals were aware of Suri's lifelong efforts to help poor farmers. Suri felt that rural farmers in India had been bypassed in large part by the Green Revolution and by modern agriculture. Throughout the previous decade working with Proagro, Suri and Edda's resolve and commitment to doing more to help these people who were living in such impoverished conditions had only increased.

Suri confirmed, "We always wanted to get involved in India's development, especially as poverty in the villages remained overwhelming, and conditions were so precarious. Most villagers are very poor. That is the salient fact. So we wanted to do something constructive to help improve their quality of life." This had been the mission of Proagro, and the philosophy of Bill Brown, Si Casady, Henry Wallace, and others at Pioneer for whom Suri had the greatest admiration.

Sehgal Family Foundation was officially established in 1998 in Des Moines, Iowa, and an application was prepared for a parallel entity in India. Suri didn't like the idea of having his own name used in creating a charitable trust in India. He preferred not to be in the spotlight in any of his philanthropic ventures; taking credit for helping others went against his personal ethic. Besides, he knew that any accomplishments to be made would be the result of a team effort anyway. But legal constraints in India and the US at the time automatically added his name to the trust application.

The original intention was for the foundation to act as a donor organization to other nongovernmental organizations (NGOs) doing important work to help the rural poor. Suri's idea was to invite projects with clear goals, deliverables, and timelines from credible NGOs, and then fund them to undertake the work the foundation envisioned. However, there were more than three million NGOs in India, and it was not a simple task to identify the ones that were trustworthy. The nonprofit sector in India did not have a reputation for probity or effective social leadership. Most of the NGOs were businesses that more typically siphoned away any government funds earmarked for rural development. Very few followed the principles of transparency and accountability in their daily operations—principles Suri considered essential.

Suri's professional experience had always been in the commercial sector, and he soon learned how complicated it was to embark on nonprofit work in India. Just to get started and be able to remit money into India, the foundation needed to be registered and approved by the Ministry of Home Affairs and meet the requirements of the Foreign Contribution Regulation Act. And any recipient NGO needed similar approval to accept foreign currency funds from a donor. The official bureaucracy moved at a slow pace with no sense of urgency. Finally receiving the required approvals, the S M Sehgal Foundation (Sehgal

Foundation) was established in 1999 as a public charitable trust based in Gurgaon, Haryana, with a defined mission to "strengthen community-led development initiatives to achieve positive social, economic, and environmental changes across rural India."[56]

When Suri's consulting responsibilities were completed in Germany in 2000, he and Edda found a lovely house on the Gulf Coast of Florida where there was room for the family to visit. Captiva Island became their home base. However, their work with the foundation and other business ventures kept Suri and Edda traveling.

Starting that year, the Sehgal Foundation provided seed funding to the Institute Plant Biotechnology Outreach in Belgium to work on stress tolerance technologies so badly needed by the rural farmers in developing countries. The foundation then provided a grant to ICRISAT,[57] based outside Hyderabad, for the development of elite sorghum and millet germplasm. The seed industry in India had already benefited handsomely from ICRISAT's presence in the country as well as its contribution to the poor in other arid and semi-arid regions of the world.[58]

Sehgal Foundation was the first private foundation to carry out research work at ICRISAT, setting up an office and renting some of their land on which to conduct crop studies. Suri met with Dr. William D. Dar, the director of ICRISAT, over dinner in 2001, and agreed, on behalf of the Sehgal Foundation, to provide ICRISAT with a multimillion-dollar endowment.

Suri asked his nephew Jay to join the Sehgal Foundation that year as an IT consultant. Jay had continued to work at Proagro (by then Aventis) after the sale of the company,[59] and was living in Delhi where Aventis had centralized its IT functions—and where his parents, Kedar and Nirmal now lived. Jay had married a beautiful young woman he

---

56. http://www.smsfoundation.org/.
57. http://www.icrisat.org/.
58. ICRISAT (International Crops Research Institute for the Semi-Arid Tropics) is the most important international research institute working on so-called "orphan" crops for resource-poor farmers, and is one of fifteen members of a global research partnership called CGIAR Consortium of International Agricultural Research Centers.
59. After Proagro was purchased from Suri by AgrEvo in 1998, AgrEvo was purchased and became Aventis CropScience in 1999, and Aventis was acquired by Bayer in 2001 (for 6.63 billion dollars). http://cornandsoybeandigest.com/bayer-acquires-aventis-cropscience.

met early in his career at Proagro, and they had two young sons attending the international school.

Jay's wife Veena had started working at Proagro at age nineteen as the secretary for the rice research coordinator. She and Jay had married in 1995 in Hyderabad on a date chosen to make sure Suri and Edda would be in India and could be present. Jay planned to move his family back to Des Moines eventually.

Both of Jay's parents needed closer attention and support when his father became quite ill. Kedar's asthma had become so severe that he was hospitalized for six weeks, and Jay's mother was no longer very mobile due to knee problems. Veena became her in-laws' main helper. She kept track of their financial and medical issues, visited with them regularly, made sure household chores were done, and took them to doctors when needed. Her caring and sincere affection for Jay's parents brought them all closer together—a blessing for Jay, who had never before had a warm relationship with his parents.

When Jay came to work full time at the Sehgal Foundation, Veena began volunteering there, working alongside the promotional people and helping to set up support programs. The head of the foundation at that time was Arvind Bahl, who was married to Suri's niece Rita (Savitri and Brij Anand's youngest daughter). Bahl had previously worked at Proagro as director of administration and human resources.

Several other projects were funded as part of Suri and Edda's larger philanthropic vision, which included rural development, crop improvement, biodiversity, and conservation. One project provided seed money for the creation of the Dharma Vana Arboretum[60] to preserve threatened and fast-disappearing historic ecologies. Its ambitious mission is to collect and preserve endangered species of trees and plants of the Deccan Plateau on 4,500 acres on the outskirts of Hyderabad, Andhra Pradesh.

The foundation also provided a large grant to Ashoka Trust for Research in Ecology and the Environment (ATREE).[61] The environmental organization, based in Bangalore, Karnatka, focuses on

---

60. http://www.arboretum.org.in/.
61. http://www.atree.org/research/sscbc.

conservation and protecting biodiversity in the Western Ghats and Eastern Himalayas. By monitoring and protecting "hot spots" in the eastern Himalayas, ATREE was tasked with building a critical body of knowledge about India's biodiversity and ecosystems in the context of global, regional, and local changes and challenges.

As the Sehgal Foundation became tied to a growing network of visionary organizations reclaiming the agricultural and ecological resources of India, a special effort back in the US was particularly close to Suri's heart. He founded, both personally and through the Sehgal Family Foundation, the William L. Brown Center for Economic Botany at the Missouri Botanical Garden in St. Louis, Missouri, to honor his mentor's interest in conservation and biodiversity. The Center's mission to "study, characterize, and conserve useful plants and associated traditional knowledge for a sustainable future" was a fitting memorial to an important visionary.[62]

Another large grant was made to Trees for Life International, a nonprofit organization based in Wichita, Kansas, with a mission to end poverty and hunger, and care of the earth.[63] This group was instrumental in planting fruit trees in India.

To further promote the spirit of giving among nonresident Indians and provide a platform for the exchange of ideas with other nonprofit organizations doing work in India, Suri helped to establish the India Development Coalition of America. This brought together organizations and individuals in the US interested in the development of India.[64]

Yet another focus of Suri's attention, starting in 2001, was directed toward a commercial venture: a seed company based in Cairo, Egypt, called Misr Hytech. The company was originally established in 1993 with multiple shareholders from several countries. Suri's family company, now called Global Investors, owned 12.5 percent at that time. But by 2001 Misr Hytech was in poor shape due to mismanagement. Various changes had occurred in investors. The company had a negative cash flow, bad debt on its books, low employee morale, and an uncertain future.

---

62. http://www.missouribotanicalgarden.org/media/fact-pages/william-l.-brown-center.aspx.
63. http://www.treesforlife.org/.
64. http://idc-america.org/.

If the company went bankrupt, seventy or eighty employees would lose their jobs through no fault of their own. That threat alone was a concern to Suri but, on top of that, he believed in the potential of Misr Hytech. So Suri decided that Global Investors would acquire all shares from anyone who wanted out of Misr Hytech, and he took control of the company in September that year. He assured the employees that everyone's job was secure and that a bright future was ahead for the company.

Suri began by brainstorming with staff in Cairo to collectively develop an action plan to turn the company around. He was determined to streamline every function. The first task was to change the business environment from an "us vs. them" dynamic to a work culture based on honesty, integrity, trust, and teamwork. The office atmosphere was made more work-friendly by involving people in decision making. The goal was to create proprietary products, innovative sales policies, production efficiencies, better collections from sales, and good financial management. The most urgent need was to create transparency and enforce ethics and integrity at all levels in the company. By employing his already proven success strategy, with slow correction and steady intervention, Misr Hytech became tight on principles but flexible in operations.

Vicki's boyfriend Ryan had finished his computer science degree at Grand View and was working in Des Moines. Suri asked Ryan if he would like to help set up IT operations for Misr Hytech in Cairo. Jay was helping there, too. Suri wanted to update and make more efficient the entire IT process at Misr Hytech, just as Jay had done for Proagro. In the fall of 2001 Ryan spent three weeks in Egypt helping to upgrade the server, acquiring and setting up new desktop computers, and installing the accounting package Jay had used at Proagro. Like Suri, Ryan would go back and forth to Cairo several times in the next couple of years for a few weeks at a time, and to India to provide similar assistance to the Sehgal Foundation.

In early 2002 Suri called on his nephew Raman, inviting him to Egypt to take over as managing director of Misr Hytech. Like Jay, Raman had also stayed on at Proagro. In 2000 he had been transferred to the Philippines by Aventis, and he and his family relocated there. His responsibilities included encouraging rice farmers in the Philippines to switch from open pollinated varieties to hybrid rice. The hybrid seed was developed by Hybrid Rice International, the seed company that Suri originally launched within the Proagro Group in India with the help of Gensuke Tokoro, to boost the incomes of rice farmers as hybrid corn had done for corn farmers.[65]

Raman left Aventis on amicable terms and took over the direction of Misr Hytech with fervor in July 2002. To bring costs in line with revenues at Misr Hytech, he paid serious attention to cash flow, cash management, and costs. Raman's experience as a seedsman with Proagro and his knowledge of each function (combined with patience, persistence, pragmatism, planning, and analytical skills) would pay off once again under Suri's direction.

Raman made several changes based on Suri's guidance and experience that he credits as the elements that led to the company's success. Low-margin nonproprietary products were discontinued. Systems and procedures were put in place for sales and product deliveries. Information technology was strengthened. Most cash transactions were eliminated with the move to a centralized banking system. The company adopted a policy of cash-and-carry sales and no credit. No sale was considered final until the seed was in the ground. At the same time, sufficient capital was injected into the company with loans, and by Global Investors, to get the fundamentals right and reduce dependence on banks to meet cash-flow needs.

Research was given top priority. Product development in corn, grain sorghum, forage sorghum, and squash was strengthened. Misr Hytech research had been conducted since the company's beginnings by a unit of young breeders under the direction of Dr. Mohamed Nasr. Suri reinforced the research by acquiring a new farm at Kanater where all research work could be carried out. This made Misr Hytech the only company in Egypt to have its own research facility. Another research station was established on leased land in the southern part of Egypt. Management provided whatever facilities were needed for research. As with Proagro, interactions were strengthened with breeders from abroad.

---

65. Suri and Gensuke still worked and traveled together, and they remained good friends. They had both fared well when the companies within Proagro were divested in 1998.

The company had begun with only one proprietary product, called hybrid 2010, registered for sale, and it had not yet proven itself in the marketplace. Everything else in corn was nonproprietary. Misr Hytech had to be creative in positioning hybrid 2010 to be successful. With good work by the breeders, the pipeline of new and better hybrids was full, putting Misr Hytech in a position to roll out new products as soon as they were officially registered, such as 2030 and 2031 among white hybrids, and Shams and 2055 among yellow hybrids.[66]

By the end of 2003 most of the fundamentals of the business were in place at Misr Hytech, and management took great care to groom and train backup staff. Employee morale was up, along with their determination to be number one in the seed business in Egypt. With the exception of bringing in Raman, and one new IT person, no changes in the management had been implemented. Changing the company culture and producing new products made the difference in the company's remarkable turnaround and subsequent growth.

The development of Misr Hytech had moved along in tandem with the changes occurring with the Sehgal Foundation. Suri had been highly frustrated with the barriers and limits the foundation faced as a donor-only organization. He knew that putting working programs into action was the proof of value, and far more effective than a lot of talk or just writing checks. He wanted to see a return on the investments, and being a donor was much too slow an endeavor.

Suri's time-honored and proven methodology in any enterprise he undertook was to bring together the best of the best people to meet his goals. Whenever he put good people together, good things started happening. That was his tactic during his twenty-four years at Pioneer, six years at PGS, and with his stunning success with Proagro. Sehgal Foundation would have a similarly inspired team.

Dr. M. D. Gupta had joined the Sehgal Foundation in 2002 to initiate a corn breeding program based at ICRISAT to access germplasm that was in the public domain from worldwide sources. Once foundation researchers were confident about uniformity and purity of inbred

66. Hybrids 2031 and 2055 turned out to be outstanding products that put Misr Hytech at the top of the market in later years.

lines, they began organizing field days, offering seed samples to breeders in the public, private, and nonprofit sectors. With this program, thousands of germplasm samples were, and continue to be, distributed to public- and private-sector breeders free of charge for use in their breeding programs.

Suri promoted Jay to executive director of the Sehgal Foundation in 2002, after Arvind Bahl retired. Jay had no real experience or training in rural work, so he spent the first months in his new role going into the villages every day, trying to understand the needs of the people and learning what the foundation was already doing.

A large brainstorming session was organized that fall. In addition to foundation people, other key individuals were invited from a dozen prominent nonprofit organizations already doing credible rural development work in India. The meetings were productive and filled with energy and enthusiasm. As a result, the decision was made for the foundation to become "active," rather than only promoting the activities of others.

As more progress could be envisioned by establishing effective grassroots programs focused on a specific poverty-stricken rural area, Sehgal Foundation went from a donor organization to an "implementing" organization. At the end of the meeting, a list of recommendations was developed and approved by the participants. That list became a blueprint for the foundation's work in the next years.

To begin implementation, Mewat, a region in Haryana with 431 villages, was selected for the foundation's attention. Though it was nearby and in close proximity to Delhi, the area had some of the lowest social indices in India—high infant mortality, low literacy rate, skewed male-female ratio in favor of males, lack of clean water, erratic power supply, lack of physical or electronic connectivity—and the region was typically overlooked by government programs already in existence.

As Suri and his team saw it, the poverty in Mewat, as elsewhere in rural India, could easily be traced to two basic problems, large family size (average 7.5) and low agricultural productivity. Land was, and still is, the only asset for most villagers, and no industry existed in the vicinity that could employ youth and provide families with supplementary income.

Suri explained, "We reasoned among ourselves that if we could help families reduce their family size by educating adolescent girls on reproductive health, and simultaneously help enhance family income by increasing agricultural productivity and connectivity with the market, we could get the development cycle going."

As was now usual, Suri and Edda were on site at the Sehgal Foundation offices for a few weeks every spring and fall. They spent time in the villages and got to know everyone on the team. When Suri wasn't on site, he was in contact with Jay by phone, especially in the beginning stages of the new initiative to make sure the foundation was on the right track with its mission and strategy.

As with every other venture Suri embarked on, an important goal was to establish the Sehgal Foundation as credible. Villagers were used to people coming in and promising things, but actually doing very little. The team had to prove their trustworthiness to the Meo people.[67]

From years of neglect and broken promises, the people of Mewat were not particularly welcoming to outsiders, including the government and NGOs that in the past had been ineffective at best and often exploitative. Transparency was essential, and building trust was a slow process. Jay and his team visited the villages and spoke to people in their homes. Team members included a medical doctor, an agricultural specialist, a social scientist, and a community mobilizer: an elderly man from one of the Mewat villages, who understood village dynamics. The team went street to street, talking to people and shaking hands.

Jay explained, "If the villagers were sitting on the floor, we sat on the floor with them. If we were offered water, we drank it. If we were offered buttermilk, we drank it. Everyone called me Jay. Showing respect in these simple ways was very important—to be accepted, to begin to build trust."

Jay learned quickly to rely on the wisdom of the community mobilizer. After noticing the trash and garbage that was strewn throughout the village, Jay decided to supply the community with trash bins on every block. On his next visit to that village, he found the children using the trash bins as swings. The community mobilizer explained that the solutions to any problem had to be "end to end" to make sense,

67. Meos are the predominant inhabitants of Mewat.

meaning that the villagers needed to have a buy-in to care about the issue in the first place; they needed to understand all aspects of any solution offered, be part of the decision-making process, and participate in the implementation of any outcome. The foundation needed to work alongside the community rather than try to form them. This insight was consistent with Suri's belief that all successful ventures needed to be partnerships. The foundation team needed to join with the villagers in all endeavors for them to have ownership in the results. In each of the villages where the programs were launched, field staff continued discussions with community members and did their best to troubleshoot villagers' concerns and suspicions.

To promote reproductive health, female staff members put a lot of effort into empowering adolescent girls through a "life skills" education program, with emphasis on building self-confidence and encouraging them to delay marriage and childbirth. Because family planning was, and still is, taboo in the predominantly conservative Muslim culture of Mewat, the team couldn't talk to the girls directly about things like birth control. Where sex education is taboo, such topics could only be addressed in a roundabout away. Sessions were held with parents and community leaders to help the villagers understand the connection between the size of families and their poverty. Separate sessions were held with boys. Everyone present was encouraged to share their viewpoints. The topic was not embraced by the male decision makers, but the women could easily see the connections. The need to find ways to empower girls and women became more and more obvious.

For income enhancement, the foundation helped farmers plant high-value crops and form cooperatives linked directly to the market, bypassing the middlemen. The foundation established linkages with the largest distributor of milk, fruits, and vegetables in Delhi (Mother Dairy) so that they would pick up produce directly from the rural farmers instead of third parties. Most small farmers only had a couple of cows, not a herd. The model had been used for milk in other areas, but this practice had never included the farmers in Mewat. Establishing the linkage made a huge difference to the income of each Meo farmer. These were results they could see.

Even with the success of these initiatives, progress was slow, that is, until the team understood that the priorities of the Mewat community were somewhat different from the priorities identified for them (a lesson relearned). By listening to the community, the team learned that the villagers' most urgent priorities were clean drinking water and healthcare. Villagers didn't care to hear about family planning or even agricultural productivity when a child in the family was seriously ill. More immediate concerns were: "How can I save my dying child?" and "Where can we find clean drinking water?"

The foundation team learned to appreciate the complexity of the development process that included multiple social, cultural, and economic factors. Issues overlapped and were intertwined, such as with water, education, and health. For example, in a village with little water, girls were less likely to go to school because they were required to spend so much time fetching water. Since more than 80 percent of the infectious diseases in India are waterborne, wastewater management was essential to prevent the spread of infectious diseases.

Recognizing that the needs of the village were interdependent, a shift was made from what the foundation team originally termed a "needs-based approach" to what they now called a "service-delivery approach." The new model was designed to integrate all aspects of village development. Projects would depend upon available funds and were intended to be replicable and sustainable.

The integrated, sustainable, village development model had four programs: water management, income enhancement through sustainable agriculture, family life education and women's empowerment, and preventive health with emphasis on sanitation and hygiene. Those four programs were to be implemented within four designated focus villages, with the most successful and sustainable programs eventually expanded to other villages. Program leaders, experts in their fields based at the foundation headquarters in Gurgaon, served as a resource group for implementation teams based in the villages, speaking in the same dialect as the villagers to further assist in communication and enhance trust within the communities.

Suri knew also that when people pay out of their own pockets, they become closely associated with what works. Villagers had to participate in implementing any project undertaken on their behalf. To that end, the villagers would be required to contribute a small part of the cost of each project, about 20 percent in cash or in-kind. The Mewat community agreed to do both.

But the villagers asked a reasonable question, "How do we know you are not going to take our money and disappear?"

The foundation team assured villagers that whatever money they contributed toward any project would not be used for any other purpose, and that all such money would be set aside for the future of that project in their village. To demonstrate this assurance, a village-level institution was set up to handle all the finances to make sure the money was tracked. All accounting was discussed with the community, and wall-sized paintings were hung in a central location to illustrate the process on a large balance sheet.

There was always some resistance in the community from those who were distrustful, or from troublemakers who sometimes created problems. Every village had those who felt that their power was being threatened. By truly listening to the community, and respecting each person's opinions, interactions improved. And by continuing to stress by example that the team was only there to help develop the village, strides were made in normalizing working relationships. Foundation team members set an example of hard work, showing up very early in summer before the heat of the day and working alongside the villagers late into the evenings as needed on their projects.

As the programs were implemented, important strategic decisions were continually made to adjust services and program components to make a greater impact. Any part of the flexible model could be addressed with greater emphasis, depending upon the most urgent needs of the village.

Another key strategy for winning the trust of the villagers was to underpromise and overdeliver. As each project was launched and people could see the results, villagers' trust in the foundation grew.

Only one village was unwilling to go along with a proposed partnership with the foundation. In a situation where two groups in the village were in opposition with each other (a common dynamic in many villages), and the foundation selected one group's idea, the other group felt resentment and refused to cooperate. When no agreement was possible, the team stepped back completely, walked away, working instead with other, willing communities. The village soon realized what they had lost and invited the foundation to return.

Since there was no sewer system in any of the villages, a model to improve sanitation, hygiene, and health included the creation of soak wells, soak pits, and inexpensive latrines (toilets). For homes with enough space, a five-foot-deep soak pit was dug and filled with rock and sand for natural filtration and safe disposal of any wastewater from laundry and bathing. This way, wastewater did not run into the streets. The team created a trap to catch soap, grease, and sediments that had to be cleaned every few days to prevent clogs in the soak pit.

In places where there was no room in houses for soak pits, a recharge well (soak well) was dug sufficient for use by twenty homes. Fifty feet deep, the soak well filtered rainwater from streets and other open spaces to help restore the water table in the village. Inexpensive latrines helped prevent the unpleasant and unhealthy conditions caused by open defecation.

Mewat is a very dry area with poor rainfall, a yearly average of only 594 millimeters, which usually comes within two to three weeks between June and September. Over-extraction of water from aquifers, and insufficient recharge, had caused the underground water to be saline and brackish in many villages. In others, salinity was creeping in. Only sixty-three villages (of 431) had "sweet" water availability as of 2014. These villages are a lifeline to this region.

Suri had the pleasure of seeing his childhood interest in the movement of water turn into cost-effective interventions for water resource management. In hilly areas, check dams were strategically placed to capture gushing rainwater to feed the aquifers instead of just letting it flow past the villages. Rainwater from rooftops was harvested for household use, and other means for collecting and storing water and replenishing freshwater aquifers were launched.

For clean drinking water, foundation researchers developed a biosand filter module in conjunction with roof-water harvesting and storage that was replicated in many schools in rural villages in Mewat in partnership with donors. Suri continued to encourage further experimentation to make technology like this more widely available.

Initiating change in education practices was much more difficult than the simplicity of building a soak pit. Education for girls stopped entirely in most villages when a girl reached the age of twelve or so. Even before that, education was sparse. Being uneducated themselves, many parents in the village did not understand the value of education in the first place. Since the parents had total control over their daughters in Mewat, the parents' attitudes had to change in order to draw girls to school. Convincing parents that literacy is a valuable skill for girls to acquire took time and effort. The education implementation team, schoolteachers, and others went door to door to enroll children for school.

One success occurred when the staff held water literacy sessions with children. After learning how effective the water program could be, kids wanted to construct soak pits in their homes. They helped to publicize the program for Environment Day with songs, posters, and drama. Children were given sheets of paper and asked to write something about the environment. An eleven-year-old girl, Arsheeda, wrote in correct English, "Water is important for our life, and we should save it."

Surprised, a staff member asked Arsheeda where she learned English.

Arsheeda said she had studied in an English medium school, but her family decided she was "too grown-up for further study." They believed she should join the community of women, which meant greater seclusion and endless household chores.

Foundation team members and the village champion, a person in each village who had been trained in the foundation's program areas, visited Arsheeda's family repeatedly before successfully convincing them to allow her to go back to school. She was thrilled.

Success with one girl, like Arsheeda, was important. School attendance was the first step in encouraging young women to participate in any community function. When any girl progressed, other girls and their parents saw the change. Gradually the number of empowered girls increased.

The foundation's income-enhancement model focused on increasing agricultural productivity, and thus earnings, through better utilization of water and soil. As in other parts of India, the soil in Mewat was degraded because of continuous cultivation over many years. The team worked with farmers, showing them how soil health and crop yields could be improved by adding organic matter and following a balanced approach in the use of fertilizers, paying special attention to the use of micronutrients. Farmers willing to try the new ideas were challenged to plant only half their field with the new methods and plant the other half the way they always had.

Sure enough, the farmers were able to see the dramatic difference between the two crops. These were smart people who saw that better agricultural practices made sense and translated to greater income.

The agricultural productivity model was successfully replicated in many villages of Mewat in partnership with Mosaic India, a fertilizer company. Lots of demonstrations in the villages followed these successes, and more farmers were willing to try new practices.

Since the goal was to take these practices elsewhere in rural India, many international visitors and potential partners accompanied staff to observe firsthand how well the programs were doing. The successes resulted in a new challenge when some Muslim religious leaders began to fear that their authority over their young people was being threatened, and that the foundation might be secretly conspiring to convert village youth to Christianity.

The implementation team addressed those fears by stressing clearly to the Meos that the foundation was strictly a nonreligious and nonpolitical organization with a goal only to help them to have a better quality of life—and with respect for all religions. This problem was usually resolved to the satisfaction of the religious leaders, but it continues to flare up from time to time.

For health and education programs that were primarily the government's responsibility, the foundation served in a more supportive role as a catalyst, assisting the public sector in their development agenda. As catalysts, the team helped to revitalize the government-mandated, but now defunct, village education committees (school boards) and worked with them to carry out small interventions to fill gaps in services, such as providing a teacher when one was not yet appointed, building latrines for boys or girls, or creating a drinking water facility. School attendance, especially for girls, went up in the villages where these things were accomplished. For the government's free school lunch program, the small effort to build a platform where food could be hygienically cooked made a big difference.

Even though most villages had a primary healthcare center, its functioning typically was poor due to complacency on the part of government officials. In an effort to make the rural clinics responsive to the needs of the community, the physician staff representative worked closely with the Department of Health to make sure that every child was immunized. One challenge was the Meos' fear that immunization was actually a sterilization program. The team physician clarified the misperception by meeting with community members. In discussing the relationship between immunization and reduced infant mortality, Meos were made aware of the intent of the program. A second physician joined the team, and life skills training classes about immunization were presented to the community in a campaign over several days. Though some people remained skeptical, the majority participated in getting the immunizations.

In collaboration with a Shroff Charitable Eye Hospital, two "eye camps" were held for people from several villages. Villagers were examined for cataracts. Those with cataracts were admitted overnight to the hospital in Delhi for surgery. Even though the city was not very far from Mewat, travel took three hours each way. The foundation paid for transportation to the hospital in Delhi, and the surgeries were paid for by an international fundraising campaign supported by the hospital.

By 2005 the Sehgal Foundation had more than 100 individual activities going on, with only a handful of people. Progress was happening in front of Suri's eyes, yet he found himself frustrated again when trying to convince potential donors of the value of the integrated, sustainable, village development model.

Suri had seen firsthand that grassroots actions were possible. He saw the hope that a simple thing like sewing lessons brought to a young girl in rural Gurgaon who could now earn income, and the difference good agricultural practices made to a poor farmer's income. He saw what clean and adequate water meant to a village, and the empowerment that health education gave to young mothers. He and Edda wanted to do more to demonstrate these successes.

By doing so many things in so many areas, it was hard to know which programs were having the greatest impact. The best results so far had been associated with water and agriculture. Success in healthcare initiatives and education were much harder to measure. The number of girls who participated in life skills training could be counted, but there was no way to quantify how each person's life had changed by achieving awareness in various areas.

Suri decided that the foundation needed an active rural research center where impact could be measured. He wanted to develop a knowledge institute, a research base, documenting the kinds of conflicts the team had encountered and how they successfully resolved what seemed like intractable problems. This could be of use to other organizations trying to help the rural poor, without having to start from scratch to learn everything the foundation team had learned from its mistakes and its successes.

If it was possible to identify the areas where the greatest impact was being made, the foundation could focus on those areas, use its money to the greatest effect, and demonstrate the value of the programs so they could be replicated elsewhere. Suri was looking at the big picture, wanting to make a difference in the most cost-effective way.

The time had come for another brainstorming session.

CHAPTER 19

Assuring Sustainability

Suri invited several of the talented people he had worked with in the past, along with social scientists, anthropologists, and experts in agricultural development and water resources to gather together with foundation staff and community leaders from the villages of Mewat to create and build a dynamic new entity for rural research and development.

The new team, some of whom came from Proagro, now focused their energy and skills on providing practical solutions for India's hundreds of thousands of rural villages that had yet to join the modern world. Emphasis had to be on programs making the greatest impact. The new entity would be housed in an ecofriendly building built to meet platinum LEED (Leadership in Energy and Environmental Design) certification requirements, be equipped with solar power that could generate enough energy for the needs of the facility, and have underground water tanks to store rainwater harvested from the rooftops. The building had to meet the highest standards of efficiency and sustainability by the US Green Building Council, meaning that all uses of energy, materials, and water met international standards for recycling, preservation, and energy conservation. The building project began in 2005 and was expected to be completed in three years. During this time, Suri tried to obtain permission to change the name of the Sehgal Foundation to Institute for Rural Research and Development (IRRAD).

That was the only part of the plan that didn't quite work out as Suri had hoped, due to technicalities in India's Foreign Contribution Regulation Act.[68]

As the Sehgal Foundation continued to carry out its programs, important legislation was enacted in India in 2005 that made a huge difference in obtaining answers from bureaucrats, which was previously impossible. This Right to Information Act also gave the foundation an opportunity to make a much stronger impact.

A serious challenge the foundation had faced all along was corruption and waste of resources by the official bureaucracy. Many years before, former prime minister, Rajiv Gandhi, estimated that only 15 percent of what the government spends on social programs ever reached the intended beneficiaries. Several laws had been enacted over the years to help the rural poor, who make up about 60 percent of India's population. The Integrated Child Development Services was enacted in 1975 to promote child health and nutrition; however, the poorer communities didn't have equal access to these services. The Public Distribution System had, since 1997, provided every village with a government ration shop to distribute food grains and fuel to the poor at highly subsidized prices. But this entitlement program was basically defunct, barely functioning due to misappropriation and lack of oversight. Embezzling was common. The people in charge took the good materials provided and sold them elsewhere at a profit. The low-quality items were given out to poor people who had no recourse.

With the new Right to Information Act, for the very first time villagers in Mewat or anywhere else could demand to know why basic government-authorized services were refused to them. The unscrupulous administrators and others who had been scamming the system and keeping people from their mandated services could no longer hide behind the walls of bureaucracy. This law put teeth into the foundation's efforts to empower the people of Mewat.

But the process to generate growing civic awareness and responsibility was very slow. The long-term community passivity and inertia

---

68. Branding as IRRAD (an initiative of the S M Sehgal Foundation) lasted for several years before the organization opted to simply use Sehgal Foundation as its brand name.

of isolation and neglect had to be transformed and mobilized into civic action in order to empower the community to demand honest implementation of the government programs available to them. The team produced posters and paintings that were hung in the villages that illustrated the appropriate allotments and entitlements.

A Midday Meal for School Children program had been initiated in 2001 by the government to provide free lunches to school children, but the meals served to the poor had become so unnourishing and sub-standard that they were inedible. The foundation posters and paintings spelled out exactly what was to be included in the meal plan.

Similarly, school principals and administrators frequently embez-zled the money allocated by the government for education facilities and services. The team taught the villagers to demand that their school committees function properly. By creating awareness in the villagers, the birth pangs of activism were slowly taking hold. Villagers began to ask why they were not given the items and services they were entitled to. People started demanding the correct amounts and quality of goods. Those in charge had to start delivering what they were supposed to deliver.

The community was participating in the creation of good gover-nance at a grassroots level. However, the empowerment process was not a fast or painless process by any means in a culture with such a deep social divide. When first approached by newly empowered citizens, government officials balked and shunned the villagers. Local bureau-crats were used to being answerable only to government departments, not poor people. Mass response from villagers was needed to mobilize change.

This dynamic steered Suri and the foundation team to a new stra-tegic approach to its work—the premise being that overcoming poverty is a matter of human rights. A focus on rights would ensure the sustain-ability of interventions at the grassroots level. The foundation moved from the service-based approach to what the team now termed as a "rights-based" approach. Emphasizing empowerment and accountability over charity fit nicely with Suri's view that people must be responsible for their own development to achieve lasting change.

As part of the rights-based approach, the "Good Governance Now" program was conceptualized by the policy and advocacy group leader and adopted by the foundation just as the new building was nearing completion. A formalized program was being prepared to create awareness on the part of villagers of their rights, with the belief that knowledge is power: once people are knowledgeable, they will be empowered to demand their rights.

By 2008 Sehgal Foundation was restructured to most effectively incorporate the most recent learnings and strategies. Several new people were hired and the foundation was organized into four centers to address the key functions and programs: Rural Research, Water Management (including agriculture), Policy and Governance, and Capacity Building. Support services for the centers included finance and administration, communication, and resource mobilization.

The Sehgal Foundation plant breeding research efforts had continued throughout this time at their leased facilities at ICRISAT. The corn breeding program that began in 2002 had expanded to include sorghum, millet, and sunflower, but sorghum and millet were discontinued later as they duplicated ICRISAT's work on those crops.

In 2004 Suri incorporated a new for-profit company, Hytech Seed India, based in Hyderabad. Suri hoped that, once he could operationalize the new company, its profits could help fund further foundation work. With this in mind, Global Investors activated the company in 2007, with an office at ICRISAT. [69] Hytech Seed India focused on developing unique germplasm and products for India and neighboring countries. A team of young scientists at Hytech Seed India experimented with corn, sorghum, millet, and sunflower.

The two seed companies, Misr Hytech and Hytech Seed India, worked closely from the beginning and collaborated in breeding and research. People from Hytech Seed India began visiting Misr Hytech regularly, and vice versa. They also traveled to participate in scientific meetings. Some Misr Hytech employees, whose mother tongue was Arabic, even went to Hytech Seed India to improve their English

---

69. The Hytech Seed India research office was moved from ICRISAT to the Medchal area on the outskirts of Hyderabad in 2012. The commercial operation remained in the city.

language skills. Hytech Seed India launched its first hybrids in sorghum, sunflower, and corn in 2008, the same year Misr Hytech paid its first dividends to investors.

The time had come for celebration on several fronts. The Good Governance Now program was ready for implementation, and the new Sehgal Foundation facility was completed on schedule. A formal inauguration celebration was held in December 2008.

The assembled crowd included the governor of the state of Haryana and other dignitaries, key funding partners (Mosaic, KMG, and the Mewat Development Agency), the staff, friends of the institution, and Sehgal family members. This was Kenny's first trip to India. Vicki and Ryan, now married, brought their little daughter Sabina, who had taken her first steps during the plane ride to India. Oliver and Ben made the trip, as did Raman and Chander.

The crowd gathered for presentations in the building's auditorium with its beautiful teak woodwork. Suri and Edda and other family members, plus a few relatives from India, were all in attendance as the program began with a choir of village girls singing the national anthem. For girls to participate in a public function was important in itself. After the welcome from Jay, words from Suri, and the presentations, a gala lunch was served, and people explored the new Platinum LEED-certified building.

Suri and Edda circulated and greeted the guests. As was typical, Suri did not wish to be in the spotlight at formal ceremonies. He instead gravitated toward a group of villagers, speaking with them informally, answering their questions and discussing their ideas and future plans. The celebration lasted all day.

At the foundation's community center in the village of Ghaghas, in Mewat, workshop tours and demonstrations were held a couple of days later. The shops were well ordered and arrayed with tool boards and equipment. In a telephone repair shop, boys were learning how to repair cellphones, and a computer workshop had teams working in groups.

Experimental agricultural rows in wooden beds, including squash, mustard, tomato, and onion, illustrated crop diversification. Vermicompost was turned over and mixed with soil to improve soil health and significantly increase crop yield. Each crop had water drips for maximum use of the precious supply.

Outside Ghaghas, visitors were shown a check dam with its ridged stone outcrops dug into the earth at many levels up and down the eroded hills, and the newly dug reservoir where runoff water was directed in the rainy season. Guests were surrounded by chattering children. Boys and men from the village surrounded and regarded the group with curiosity, listening to the descriptions of the water program, including collection, storage, ground water replenishment, recharging of wells, deepening of ponds, and roof-water harvesting.

At Notki village, brightly colored signs explained the foundation programs going on there. Education team leaders made brief presentations, pointing out village participation and the funds they had raised. Villagers received Edda and Suri by placing an elaborate and colorful turban on Suri's head, designating him a high benefactor. A silk shawl was put around Edda's shoulders.

Guests were directed across the schoolyard into a classroom where girls wearing headscarves were seated in rows to one side. Suri addressed the children and their teachers in Hindi, asking them to volunteer descriptions of what they were learning, and listening to their animated answers.

The villagers appeared comfortable in a newly acquired sense of order. Students and workers spoke with enthusiasm and assurance, volunteering descriptions of the projects they were part of. An orchard had been planted as a means of preventing desertification and to create an income for the panchayat (local governing body akin to a town council). Wells had been dug, with new pipes and storage tanks. Guests were shown the modern well head and pump, soak wells, and soak pits. The street was lit with solar lampposts, and the main street was paved with bricks. The health center included birthing rooms with modern equipment and screens for privacy, plus wall diagrams showing medical and preventive health procedures.

There was still plenty of work to be done, as evidenced by the women of the town, who watched everything from behind their houses at a little distance across the street.

The most important and effective models were finally identified and became the basis of the foundation's focused approach to social change: water management, agricultural development, and good governance. Though the foundation's literature did not openly address gender equality, each model was directly related to the empowerment of women. The team had always known that women suffer the most when water is not available. For girls to be forced to abandon education in favor of fetching water for the family and for crops (since 70 percent of the water is used in agriculture) perpetuates their deeply entrenched subservient role in Indian society; and the team's experience had proven that empowering one woman changed the dynamic of her entire family, which in turn affected the rest of the community. Empowering women would drive village development. So the team's mandate was to make sure that every training workshop had at least 50 percent women participants.

People who had gone through the training workshops felt increasingly empowered to demand what was due to them from the bureaucrats. Other villagers saw the benefits and began to participate in the democratic process. Empowering rural communities to take ownership of their own development assured the sustainability of the good-governance model and other rights-based initiatives designed to provide the rural poor with a path to human dignity, justice, and vertical mobility.

With the new foundation facility humming along, and the new rights-based approach being successfully operationalized by the hard-working team in a half-dozen villages, Suri was still challenged to find more powerful ways to make a bigger difference to more people. He was not satisfied with several programs operating successfully in a few test villages. He wanted to achieve strategic depth in more villages.

The Sehgal Foundation's financial resources could cover the costs of implementing the most effective service-delivery models in a cluster of twenty villages with a community center in close proximity. But one cluster of twenty villages out of about 650,000 villages in India was a mere drop in the ocean. Scaling up was recognized as a must for a larger impact; the foundation had to raise additional funds from other sources.

As part of scaling up, Jay and his family moved back to Des Moines where he began to represent the foundation to American donor organizations and development institutions. His task was to mobilize resources from donors, create linkages with key research organizations and foundations, and attract volunteers and postdoc candidates from the US to work at the foundation to gain exposure to rural India, and hopefully continue to pursue their interest in rural India when they returned home.

Publications and videos made available on the website were designed to demonstrate how programs in the field actually worked and invited international contacts, students, and researchers to join the Sehgal Foundation and participate in future activities. The foundation launched a community radio station that was broadcast in the local language to the villages in Mewat. Local-language print media helped to increase awareness of citizens' rights. Conferences and consultations on governance were presented, providing platforms for discussions among villagers, government officials, NGOs, and others.

As citizen participation in the democratic process increased, several village panchayats learned how to test state and federal administrations, standing on their constitutional rights to public information, proposing solutions from their own knowledge and demonstrated competence, and demanding action. Villages the team worked with slowly began to establish reciprocal relationships with government officials for addressing mutual responsibilities. They served as the model for greater cooperation between government and local groups able to formulate agendas for change.

Good Governance Now continued to spread to villages throughout the Mewat region over the next few years. Working in partnership with other individuals and organizations, the Sehgal Foundation's long-term goal continued to focus on converting the Good Governance Now model into an all-India movement.

By 2010 a dramatic shift occurred in the Public Distribution System program in the foundation villages that resulted in more people receiving the goods they were entitled to. For creating the awareness of these entitlements within the villages, the Sehgal Foundation heard some loud objections from many who had previously benefited from the long-entrenched mismanagement.

With one NGO for every 400 people in India, Suri envisioned that effective reform in this sector would have a dynamic impact. He knew that if Indian NGOs would stop squandering resources and betraying the people they were tasked to serve, and instead become credible by being scrupulously transparent and accountable in their daily operations, it would profoundly change India.

In the absence of a mechanism for identifying or classifying trustworthy NGOs, Suri wanted to make sure that the Sehgal Foundation remained highly credible. From its start, the foundation maintained a strict policy against bribes or dishonesty; integrity at every level was a hallmark of the organization. With a growing national movement against government corruption in India and the call for transparency in all fiscal affairs, the foundation was (and remains) a vanguard institution. Presenting conferences each year on water management and good governance offers a good platform for the discussion of these issues with representatives of NGOs who are invited to participate.

Jane Schukoske, the CEO of the Sehgal Foundation since 2011, immediately pushed for placing even more of the foundation's focus on the empowerment of women; for the first time the Sehgal Foundation's gender equality policies were put in print, including emphasis on programs led by and responsive to the needs of women.

Jane's expertise and her contacts with law schools in India have been a perfect fit for the further development of rights-based strategies. She had previously taught at the Baltimore School of Law and had worked for years with the Fulbright Program in India, sending scholars from India to the US. She has been able to bring villagers into contact with law students and professors to refine the foundation's rights-based model. With the Good Governance Now model embraced by all villages in the Mewat District as of 2014, the approach is on track to become a nationwide movement.

Shortly before Suri's eightieth birthday in April 2014, the Corporate Social Responsibility (CSR) Policy Rules under India's new Companies Act went into effect, which mandated that two percent of net business profits from large companies in India, including foreign companies, would go to CSR initiatives.[70] This mandate will no doubt be a useful tool in the Sehgal Foundation's arsenal for multiplying the impact of its sustainable programs.

Suri's vision is articulated in the Sehgal Foundation goal: with the appropriate support from civil society organizations and the government, rural communities will not only be capable of transforming awareness into action, but will be mobilized to develop their own vision for development, ultimately leading to a more dignified life.

70. http://www.mondaq.com/india/x/305620/Corporate+Commercial+Law/
Corporate+Social+Responsibility+Mandating+Companies+To+Contribute+Towards+Society.

CHAPTER 20

Blessings

In the spring of 2011 Suri and Edda were in Japan when a sudden crisis unfolded. Edda was with nephew Chander and his family in the countryside, and Suri was in Tokyo, visiting his old friend, Gensuke Tokoro. This time—unlike some past episodes, such as wars, revolution in the Dominican Republic, or the rebel uprising in Nicaragua—the cause of upheaval around Suri and Edda was "natural" rather than political.

Suri was in a meeting with Gensuke and a few others in his sixth-floor office when the building suddenly started shaking. Pictures fell off the walls and books plummeted from shelves. Though realizing it was an earthquake, they didn't worry too much because minor quakes were not unusual in Japan. But when the water started spilling from the fish tanks, everyone became very concerned. Luckily the tremors stopped and people could immediately vacate the building. But Suri and Gensuke had an important meeting that afternoon with the chairman of the Nippon Foundation, Mr. Yohei Sasakawa, in his office nearby. Nippon and its affiliates, Tokyo Foundation and Sasakawa Peace Foundation, together constitute the largest nonprofit foundation in Japan. The meeting was to explore the possibilities of cooperative efforts between the Sehgal Foundation and their foundation. Unfortunately, the very productive meeting came to an abrupt halt when sirens went off, warning them to vacate the building immediately due to danger from earthquake aftershocks.

Gensuke had planned an event for that evening at a private club. Every detail was arranged for Suri to make a presentation about the Sehgal Foundation to potential donors, followed by a dinner party. The event was cancelled due to the unfolding tragedy, but there was no way to inform the fifty-plus guests Gensuke had invited.

Eleven people showed up, mostly friends and staff from the office. The small group enjoyed drinks and the excellent food, not realizing the extent of the devastation, especially to the region about 200 miles further north. The magnitude-9.0 earthquake, followed by the enormously destructive tsunami, would be described as one of the worst catastrophes in Japan's history.

All public transportation in Tokyo was at a standstill. At Suri's hotel, the lobby was full of people sleeping in chairs and on the floor. When he reached their room, he did not find Edda. All cell phones were down, and they had no way to contact each other.

Edda, Chander, his wife Rumiko, and his mother had been in a shopping mall forty miles from Tokyo when the quake hit. Traffic jammed the roads as people tried to rescue their stranded and displaced family members. Edda finally made it back to the hotel by 2:00 a.m.; it had taken more than ten hours to drive the forty miles. Once again, the Sehgals had emerged unscathed by the difficult circumstances around them.

Years before, they had been present in New York City on a hot day in August when a huge power outage shut down the city. Suri and Edda were touring the Metropolitan Museum of Art when the power went out all over the Northeastern US, affecting eight states and the Canadian province of Ontario.[71] No electricity in New York City meant no subways, no other railway transportation, and no air travel. People were trapped in elevators, the streets were gridlocked, and commuters were stranded. Suri and Edda were able to simply walk back to their hotel and climb the twenty-six floors to get to their hotel room for the night. For them, the blackout was merely a minor inconvenience, not worthy of complaint (though Suri recalls, ". . . except when a store nearby wanted $32 for an ordinary flashlight").

---

71. The Northeast blackout of 2003 affected more than 55,000,000 people in the US and Canada. Full power was not restored for two days.

Suri and Edda have felt continually blessed throughout their lives together. Though they both had to navigate profound dangers, hardships, and uncertainties from early childhood, they each went on to face life with buoyancy and hope. Their closest associates attest that their spirit of optimism is contagious.

Though Suri and Edda have sometimes received recognitions of various sorts for their widespread philanthropy, public acknowledgment is something they do not seek or feel comfortable with, participating only to bring further attention to important social justice work being done in the US, India, and elsewhere to help the poor and to conserve biodiversity.

In June 2011 at the American India Foundation's Spring Awards Gala, Suri received a Leadership in Philanthropy Award. Edda, Suri, Kenny, Jay, Oliver, and Sehgal Foundation CEO Jane Schukoske attended the formal event held at Cipriani Wall Street in Manhattan. The invitation read, "Attire for the Gala is Black Tie or Indian Formal." Suri's accomplishments were enumerated, with emphasis on the foundation's cost-effective strategies in water management, income enhancement, hygiene and sanitation, and the empowerment of women and girls in India.

In 2014 Suri was recognized by the Indian Confederation of NGOs with a Global Indian Karmaveer Puraskaar Lifetime Achievement Award.[72] Receiving this Global Award for Social Justice and Citizen Action was an opportunity Suri used to call on other NGOs in India to join him in endorsing transparency, accountability, and good governance practices.

At eighty years old, Suri remains active in world councils with Edda by his side, supporting and directing projects in the US and abroad. The Sehgal Foundation has a consultative status with the United Nations Economic and Social Council, and Suri is a member of the Clinton Global Initiative. He continues to advise and visit with his teams of hybrid seed breeders with the Sehgal Foundation and with both seed companies, Misr Hytech in Cairo and Hytech Seed India.

---

72. http://karmaveerglobalawards.com/Awardees2013.html.

Suri's manner with the teams remains casual but focused, relaxed but probing, and always returning to the questions at hand. Suri and Edda continue to believe that people are key to the success of any enterprise: put good people together and give them responsibility and the ability to develop their skills in an honest, respectful, and supportive environment. He and Edda understand that an organization must have the necessary patience to train people who are perhaps working at different speeds with different abilities. "We must bear with their shortcomings and build on their strengths."

In all meetings, Suri insists on using the term "we" rather than "I." He explains the importance of giving credit to others: "As they say, success has many fathers; failure has none. 'We did it' is the common expression in our companies."

Suri's successful businesses and projects in the public and private sectors are proof of the merit in his methods, his philosophy, and his ethic, which Edda fully shares. Together they continue to promote the Sehgal Foundation values of "integrity, professionalism, excellence, and optimism."

Nephew Raman agrees wholeheartedly with Suri's approach to business, to philanthropy, and to living. "We have been able to ensure a good life for the people who work with us at Misr Hytech. Suri is my uncle, but he is my boss too, a leader who provides guidance. He is honest and fair with people. He trusts us, and so we trust him. Mutual trust in a company works wonders. We preach the same principles to others that we have learned from him. He has always helped the family, and we are thankful for what we have become. He gave all of us opportunities for education, and we do the same for our people. They get training, but they must learn how to work well and stand on their own feet. He teaches people to become independent yet still work in a cooperative way. He does not believe in charity. That's not his philosophy. He is patient and encourages people to get an education and/or develop a skill. We have many relatives in the US, who are medical doctors or other types of professionals, and they are very well-off—but their base has been Suri. He is a scientist cum businessman, but above all he is a wonderful human being with a soft heart."

After Proagro, Misr Hytech is one of Suri's favorite success stories. It became the leading seed company in Egypt in 2010. Farmers are benefiting from its good products, shareholders are receiving handsome dividends and, true to Suri's practice, the company's good fortune has been generously shared with all the employees year after year. Employees know that, as the company grows, they will continue to benefit, too.

The economic and political situation in Egypt has not always been easy. However, despite social and political turmoil, Misr Hytech's growth has continued. The company has established a good reputation and keeps in regular contact with the Egyptian Ministry of Agriculture, the US and Indian embassies, and the expatriate businesses in the country. As a member of the American Chamber of Commerce, the company stays in contact with other US businesses in Egypt.

As of 2014 Misr Hytech is the most successful seed company in Egypt, topmost in market share, and highly profitable, with promising opportunities in neighboring countries and the rest of Africa. Once again, motivation of the people made the difference in turning failure to triumph.

Misr Hytech supports the Sehgal Foundation's philanthropic efforts in uplifting the lives of the rural poor with agricultural projects that expand farmer awareness, assist women farmers, and help farmers to produce higher yields in sustainable ways.

Suri is confident that Hytech Seed India can follow the path of Misr Hytech, perhaps on an even larger scale once it has a good cash flow, with the potential to expand to the Philippines, Thailand, and Indonesia.

Raman adds, "Suri and Edda's efforts continue to help the less fortunate, creating a commitment among employees and staff to support the common good. The combined efforts of the Sehgal Foundation, Misr Hytech, and Hytech Seed India are an inspiring example of humane development."

Hiring members of his own extended family has been a frequent strategy throughout Suri's career, combining two vital qualities in the developing world—a strong commitment to, and an innate trust, in family. Suri explains, "Though I started with my family, in general, I trust others as well. I trust people until they prove they cannot be trusted. And I am rarely disappointed."

Suri's many good friends, colleagues, and associates around the world are grateful recipients of his trust and multipliers of his generosity.

Suri and Edda remain very proud Americans. Edda still gets teary when she hears the "Star Spangled Banner," and she and Suri are full of gratitude for the many blessings that life in the United States has brought them. From their turbulent backgrounds as refugees from displaced cultures, they found each other at a remarkable time in the history of their respective native countries and made an astonishing life together in America.

Suri reflects, "What impressed me the most when I came to the United States was the easiness of life and the kindness of the people. Coming from a complex society like India, I felt as if I landed in heaven. Here opportunities are unlimited for those who wanted to work hard and enjoy freedom. Although we can now afford to live anywhere in the world, America is still our first choice. It is our home."

Still based in Florida, Suri and Edda continue to travel around the world, keeping tabs on their family, the seed businesses, the Sehgal Foundation, and their other philanthropic efforts. They attend family meetings annually at the Global Investors headquarters in Des Moines, where the entire family gathers to discuss current projects and propose new ones for the family foundation to support.

Suri and Edda's children continue the "giving back" legacy, assisting a range of individuals and organizations making a difference locally and around the world in the areas of health, education, and conservation. Their efforts have addressed water scarcity and preservation; protection and restoration of land, water, and wildlife; and medical services for the poor in underdeveloped countries. Some projects are small and specific, while others are more global and far-reaching. But each effort is in keeping with the goal of making a positive difference in the lives of others.

Epilogue

Speaking for himself and for Edda, Suri adds the final words to this story: "Success is a relative term that means different things to different people. We have never measured success in terms of earning power or status. We always lived within our means. We were not extravagant. We have lived a simple but good, clean life. We know that money and possessions are not lasting. What lasts is the deep satisfaction of achievement and knowing that the people we served, worked with, trained, or helped are better off because we were there to lend a hand. In anyone's life there are good and bad moments; that is how life goes on. The sad moments in our life were very few, and the happy moments have been many. We always wanted to help others in need, and that was our greatest satisfaction. Any adversity that we experienced turned out to be a blessing for which we are grateful. We are grateful for so many blessings."

Afterword

The world sorely needs people like Suri and Edda Sehgal. In the eight decades since Suri's birth in 1934, the population of India has multiplied well over four times to almost 1.2 billion of the now 7.2 billion on the planet. At present the world population is collectively using about 1.6 times the sustainable productivity of the planet, with some nations consuming far more than they can produce.[73] But there is hope for this earth, and *Seeds for Change* vividly and concretely illustrates why that is so.

The inspiring saga of the lives of Suri and Edda Sehgal provides many lessons for us all, as individuals and nations continue their self-serving actions and engage in seemingly endless battles to accumulate increasingly unfair concentrations of the world's wealth for themselves. Suri and Edda each contended effectively with global problems in their own ways. The values that were so strongly manifest in pre-partition India and Suri's upbringing in an atmosphere of generosity and the acceptance of people with very different beliefs, clearly determined the course of his life. The same values match those that were originally present and ultimately prevailed in Europe, where Edda and her family found their way amid the chaos and destruction of World War II.

Given what Suri and Edda learned during their childhoods, it is no surprise that these remarkable individuals went on to forge a remarkable marriage that has already spanned a half-century. Their relationship with each other and their values are projected to the world beyond and the people in it. Their lives, standards, and actions clearly echo the highest values of pre-partition India and pre-World War II Europe. These represent in part an ideal to which we must aspire if our civilization is to prosper in the decades and centuries to come as the earth's population

---

73. http://www.footprintnetwork.org.

soars. Suri and Edda, their character and their love, pose a substantial challenge of service to the younger generations of their family, and all of our families, who will ultimately help to determine the future course of the world. Having personally observed the early decades of the lives of these younger people, I am confident they will rise to that challenge and continue to make Suri, Edda, and all the rest of us proud of them.

Peter H. Raven PhD, president emeritus
Missouri Botanical Garden

SEHGAL FOUNDATION is a true partnership between donors, volunteers, interns, employees, researchers, and rural communities.

SEHGAL FOUNDATION
100 Court Ave, Suite 211
Des Moines, Iowa 50309-2256
USA
Phone: 1 515 288 0010
email: sff-usa@smsfoundation.org

Sehgal Foundation is a US-based 501(c)(3) tax-exempt private foundation. Donations are deductible for US tax purposes as allowed by law.

S M Sehgal Foundation
(SEHGAL FOUNDATION)
Plot No.34, Sector 44, Institutional Area
Gurgaon, Haryana 122003
INDIA
Phone: 91-124-4744100
email: smsf@smsfoundation.org

S M Sehgal Foundation has 80G status in India. All donations made within India are 50% tax deductible.

With support from donors and partners, Sehgal Foundation works together with rural communities to manage water resources, increase agricultural productivity, and strengthen rural governance through citizen participation and emphasis on gender equality and women's empowerment.

Join the team and see what we can do together!

www.smsfoundation.org

Acknowledgments

The collaboration that grew into *Seeds for Change: The Lives and Work of Suri and Edda Sehgal* began about seven years before its publication in 2014, soon after the decision was made to document the lives of Suri and Edda for their family. Roy Skodnick was asked to begin collecting materials and interviewing members of the Sehgal family, colleagues, and business associates, which included travel to India to visit the S M Sehgal Foundation and see the important work being done there. In 2012 I was asked to take the information accumulated by Roy, as well as material family sources compiled in the meantime, plus new information, and create a narrative of the family story. I conducted more than 150 additional hours of interviews in person and on Skype and exchanged hundreds of emails with Suri, Edda, members of their family, and others in the US and abroad over the next two years. Help was provided along the way from teams of outside readers, family, and friends, who offered added tidbits, clarified details, and asked questions, prompting further fact checking, which uncovered new scraps of information that helped make this family story as accurate and clearly told as possible—and in time for publication in 2014 to coincide with significant anniversaries: Suri's eightieth birthday, the couple's fifty years of marriage, their forty years as proud and grateful American citizens, and the fifteenth anniversary of the establishment of the S M Sehgal Foundation in Gurgaon, India.

My sincere thanks go to Ambassador Kenneth Quinn, Dr. Peter Raven, Rekha Basu, Elder Carson, Tad Cornell, Dr. Max Cowie, Dr. William D. Dar, Janet Gage, Michael Gartner, Judge Jim Gritzner, Susan Thurston Hamerski, Mayapriya Long, Patricia Morris, Pooja Murada, Jane Schukoske, Anne Seltz, Dr. Prem Sharma, Roy Skodnick, Gensuke Tokoro, Geert Van Brandt, every member of Suri and Edda's family—and especially to Suri and Edda. Being included within the orbit surrounding them, and being a partner in the documentation of their family legacy, has been a tremendously enjoyable honor. I am profoundly heartened by Suri and Edda's generosity of spirit as they have continuously assumed responsibility in the larger world, doing what they could to make life better for the poorest of the poor and helping to assure the welfare of the planet we all share.

Marly Cornell

About The Author

Marly Cornell is an artist, writer, and social justice advocate. Her book *The Able Life of Cody Jane: Still Celebrating* (LightaLight Publications, 2011) is the winner of a Midwest Book Award. She lives with her husband Ernie Feil in Minneapolis, Minnesota. www.theablelife.com

SEEDS FOR CHANGE

The body text for *Seeds for Change* is set in Adobe Garamond® Pro, a contemporary typeface family based on the roman types of sixteenth century French type designer, Claude Garamond, and the italic types of Robert Granjon. The title and chapter numbers font is Serlio LH.